Sustainable Cities

AF811306

Claudio Scardovi

Sustainable Cities

Big Data, Artificial Intelligence and the Rise of Green, "Cy-phy" Cities

Springer

Claudio Scardovi
HOPE SpA
Milan, Italy

ISBN 978-3-030-68437-2 ISBN 978-3-030-68438-9 (eBook)
https://doi.org/10.1007/978-3-030-68438-9

© The Editor(s) (if applicable) and The Author(s), under exclusive license to Springer Nature Switzerland AG 2021
This work is subject to copyright. All rights are solely and exclusively licensed by the Publisher, whether the whole or part of the material is concerned, specifically the rights of translation, reprinting, reuse of illustrations, recitation, broadcasting, reproduction on microfilms or in any other physical way, and transmission or information storage and retrieval, electronic adaptation, computer software, or by similar or dissimilar methodology now known or hereafter developed.
The use of general descriptive names, registered names, trademarks, service marks, etc. in this publication does not imply, even in the absence of a specific statement, that such names are exempt from the relevant protective laws and regulations and therefore free for general use.
The publisher, the authors, and the editors are safe to assume that the advice and information in this book are believed to be true and accurate at the date of publication. Neither the publisher nor the authors or the editors give a warranty, expressed or implied, with respect to the material contained herein or for any errors or omissions that may have been made. The publisher remains neutral with regard to jurisdictional claims in published maps and institutional affiliations.

This Springer imprint is published by the registered company Springer Nature Switzerland AG.
The registered company address is: Gewerbestrasse 11, 6330 Cham, Switzerland

Foreword

Cities are the subject of several sci-fi novels and movies. Usually artists imagine catastrophic scenarios for urban life; just think of the chiaroscuro description in the cult movie *Blade Runner* or the submerged Manhattan in Spielberg (and Kubrick) *Artificial Intelligence*. Alternatively, novels like Clifford Simak's *City* or Arthur Clarke's *The City and the Stars* explore futures like the ones mentioned in the book with "human-kind preservation plan in caves underground, or as a long-term project to colonize Mars." It is therefore very welcome to read a structured reflection on possible futures of cities, where technology helps sustainability, and, borrowing from the book's terminology, people evolve to cyborgs but remain free to determine their own destiny.

One of the many highlights of the book is thinking about sustainability in general, a very useful enterprise. Google Books Ngram Viewer shows a constant increase between 1950 and today in the use of the acronym ESG (environment, social, governance) and an even larger growth of the term "sustainability" that, as of today, is 35 times more popular than ESG. It has become an economic narrative in the sense advocated by Yale's Nobel Prize winner Bob Shiller. But what is exactly sustainability? We are not quite sure. Claudio, however, provides his own definition: "a sustainable city is one that is not just able to survive and protect the wealth and wellbeing of its citizens through extreme events, such as pandemics, wars and extreme climate change scenarios—but also one that is democratic and with a good form of government, free and meritocratic but also socially sensitive and compassionate."

To many people, you can call it a sustainable city or you can call it Heaven. But how do we find the Stairway to Heaven? While Led Zeppelin is not useful here, the book offers several ideas, revolving, in my opinion, on two fundamental themes: data (including new computational techniques, enlarged data storage capabilities, new ways to monitor real life and turn it into a Matrix like history) and institutions (including better ways to manage companies and governments, smart rules and regulations to limit negative and to foster positive externalities, effective schools to educate people of all ages to be proficient at wealth and technology management, financial products that extend the list of shareholders of assets and cities themselves). Claudio Scardovi helps the reader in thinking about current and future developments in data and institutions, and in imagining what lies ahead. In its futuristic style and in its effort at inviting us to think of the next 10–20 years, his work reminds me of great books that I have enjoyed reading and have shaped my view of reality, like Shiller's *Irrational Exuberance* and Graeber's *Debt.*

Sustainable Cities is for everyone. Economics and finance students may find a way to apply their technical knowledge to a rigorous analysis of current trends. Novel lovers may find a fascinating writing style that talks about the way we live and produce. However, I would like to see managers, at the juncture between public and private organizations, read and use the book. Civil servants may find a new way to understand the relevance of their job and imagine ways to leverage their skills and jobs. Private managers may put their profit creation duties in perspective and better see the social purpose of what they do in an (increasing) interconnected reality. And I would like to see politicians and city administrators reflect on the goals they should set to the communities that serve. Most of all, I would like to see investors, especially institutional investors, read the book, rethink their way of defining risk and return, and view differently the final implications of their investments.

As the book suggests, big crises are big opportunities and require totally new frameworks. Important social arrangements that characterize our societies today, from the social security system to the governance of the international economy, were born out of the need to invent and change. But inertia is everywhere. Humans tend to rely on their comfort zone. Our brain is not made to spontaneously search for new solutions. In mathematics, we have used for ages local optimization rather than global optimization techniques. The book is an invitation to think of our current way of living and to reflect upon the deficiencies that COVID-19 has highlighted, and to act. It is quite possible that the level of challenges facing us requires totally new arrangements, for example in the coordination between public and private actors, which may well have to go beyond their current hide-and-seek mood to embrace joint efforts to solve joint problems.

The AC world requires a thorough intellectual and institutional framework and new benchmarks to judge what we do. In my view, thinking about *Sustainable Cities* requires how to turn cities from the current state of quasi-bankruptcy to lively sources of economic profit (not a bad word after all) in respect of all stakeholders, including our sons and their sons. Profit producing cities may be sustainable, but a loss-making city is, by definition, unsustainable. Competition forces every city to become an efficient data generation machine. We need new collective choices and private companies that exploit new data analysis to create valuable institutions and turn favelas into tech clusters and crowding into density. To sustain the span of issues, the study of mathematics and computer science must be completed by the study of ethics and psychology. The book contains phenomenal insights on how to do that.

Department of Finance, Bocconi University, Milan, Italy

Andrea Beltratti

Contents

1 From Smart to Meta Cities 1
1.1 Towards "Other Normals" 1
1.2 Second Sleep 2
1.3 If Smart Is Not Enough 4
1.4 From Smart to Cyber Cities 6
1.5 Beauty, with a Purpose 8
1.6 Cyber Cities Are Not for Cyborgs 11
1.7 Death of the Smartphone 13
1.8 From Cyber to Meta Cities 14
1.9 A "Cy-Phy" Humanism 17
References 20

2 A Cy-Phy and Fairer Economy 21
2.1 Digital Is Passé, Without Physical 21
2.2 Close Twins 22
2.3 Alice's Mirror 25
2.4 Digital and Physical-Digital 27
2.5 Socially Acceptable Hence Sustainable 28
2.6 Harnessing Data for the Common Good 30
2.7 Smart, Cyber, and Meta Cities at the Time of COVID-19 32
2.8 Fair Use, Fair Trade 34
2.9 Big, Digital, and Smart 36
2.10 Data Traders and Arbitrageurs 38
2.11 Fair Digital in Fair Cities 39
2.12 Death of Cities? 41
References 43

3 Info-telligence in the City 45
3.1 Info-telligence in the City 45
3.2 The Economics of Info-telligence 46
3.3 Data Are (Almost) Unique 48

3.4 Attention and Intelligence Utilities . . . 49
3.5 The Risk of Being "Big Digital" . . . 51
3.6 A "Physical" Missing Point . . . 53
3.7 Death of the Smartphone . . . 55
3.8 Architectural Evolution and the Redesign of the City . . . 57
3.9 Where Would the Intelligence Sit . . . 58
3.10 Geopolitical Implications . . . 60
References . . . 62

4 Sustainable Cities and Climate Change . . . 65
4.1 Getting Hot . . . 65
4.2 Sustainable Cities . . . 66
4.3 It's All About Carbon (Dioxide)? . . . 69
4.4 An Anthropological Phenomenon . . . 71
4.5 The Risk for the City . . . 73
4.6 AI and the Search for Serviceable Truths . . . 74
4.7 It's All About (Climate) Risk Management . . . 76
4.8 Green Is the Color of Money . . . 78
4.9 A Big Data Approach for Climate Change . . . 82
4.10 Green to Death . . . 86
4.11 A Carbon Tax or a Green Derivative? . . . 88
4.12 Credibility Matters . . . 90
4.13 Climate Matters . . . 92
4.14 Climate Conscience . . . 94
4.15 An Anthropological Challenge . . . 95
References . . . 96

5 Our "Other Normal" . . . 97
5.1 The Day After . . . 97
5.2 The Death (and Rebirth) of the Office . . . 99
5.3 Social (Distance) in the City . . . 101
5.4 Locking Down . . . 103
5.5 Unlocking . . . 105
5.6 Restart in a Post COVID-19 World . . . 106
5.7 Creative Destruction at the Time of the Coronavirus . . . 109
5.8 Public Good . . . 111
5.9 Finance, Real Estate, and Sustainable Cities . . . 113
References . . . 115

6 The Rise of the Cy-Phy Company . . . 117
6.1 Smart, Cyborg, and Singularity Driven . . . 117
6.2 The Advent of the Cy-Phy Bank . . . 119
6.3 Singularly Cyber-Physical Bank . . . 120
6.4 A Digital Wealth-Being Manager . . . 121
6.5 A Data Bank Is Already Here (and Is Worth Billions) . . . 124
6.6 A Star Is Born . . . 128

6.7 From Dumb Money to Dumb Cables . . . 130
6.8 Cy-Phy in the TMT Industry . . . 131
6.9 At Your (Digital) Health . . . 135
6.10 Poor Patient . . . 137
6.11 Googling "Health" . . . 140
6.12 Healthcare of One . . . 143
6.13 An eHealth Wallet for Sustainable Cities . . . 145
6.14 Strategic Forays in AI-Based Use Cases . . . 147
6.15 Serviceable Truth . . . 152
References . . . 157

7 Investing in Real Estate: BC/AC . . . 159
7.1 Real Estate BC/AC . . . 159
7.2 This Time Will Be Different . . . 160
7.3 The Journey and the Destination . . . 165
7.4 Financialization Ahead . . . 170
7.5 A Geopolitical Play . . . 173
7.6 The Role of the State in Real Estate BC . . . 174
7.7 The Role of the State in Real Estate AC . . . 178
7.8 A Private Equity Opportunity for Everybody . . . 180
7.9 From the 15 Minute City to the Integrated Building . . . 181
7.10 New Paradigms for Real Estate Investing . . . 183
References . . . 184

8 Open Data in Open Cities . . . 185
8.1 Electrifying Future . . . 185
8.2 Open Data Revolution . . . 186
8.3 Going Soft . . . 188
8.4 Open Cities, Open Companies . . . 190
8.5 The Real Risk in Front of Us . . . 192
8.6 Towards a New Model for Cities . . . 194
8.7 The Leopard on Mars . . . 196
8.8 A Bit of Clouds Could Help . . . 198
8.9 Defining Moment . . . 200
8.10 Cities as the "Cy-Phy" Platforms of the Future . . . 203
8.11 Conclusion: From Smart to Sustainable Cities . . . 204
References . . . 206

9 Finance, Real Estate, and the "Smart City" Challenge in Milan Area: Toward a Sustainable Future? . . . 207
9.1 Finance Role in the City's Development . . . 207
9.2 Big Data's Role in Real Estate . . . 208
9.3 Real Estate Transformation: From a Product-Centered to a Customer-Centered Industry . . . 209
9.4 The Sustainability Paradigm . . . 210
9.5 Milan and the Smart-City Challenge . . . 210

Appendix: Behind the High-Tech Curtains of a Sustainable City 215
SCaaS: Sustainable City as a Service and Open Data 215
Edge Computing and Augmented Reality as Personal Shoppers 216
Weather of Things to Produce Healthy Food 218
Being at Madison Square Garden While Being at Home 220
Air Delivery and Inspection 221

Chapter 1
From Smart to Meta Cities

1.1 Towards "Other Normals"

In the words of the former President of the United States Barak Obama, no good crisis should be laid to waste as an opportunity for change. Should we judge present times on the basis of the deepness and extent of crisis we are all now living in, it would look like we do have now a great opportunity for change—in the way we work, live, across most geopolitical blocks and in the context of urban areas and global cities that are now looking like as "too crowded" to ensure "social distancing," and too interconnected, across geographical zones and value chains, to easily make up for any potential supply chain disruption led, for example, by the force of nature, a global war, or a simple pandemic. Since the start of the planning and writing of this book, lots of things have changed, for the author and for the world we all live in—with expected falls in the GDP of the European Countries and of the United States expected in the high double digit, and negative growth at global level as well, with a deep recession and with an uncertain shape and behavior (as "V," "U," or "L") to follow. COVID-19-led health and humanitarian crisis has in fact brutally reminded us of the extreme fragility of our world—that we had been taking for granted with its secular trends of international of commerce and finance, for its interconnectedness and specialization based on free-trade principles and as a "just in time," lean-managed with no slack attached, super-efficient and integrated global market for goods and services. A simple virus has shaken all that and accelerated the recognition of our, far from assured, sustainability, as humankind led set of civilizations. Pandemic has been taking us quite unexpectedly and wholly unprepared and is forcing entire countries and their main cities to rethink their future ways in likely radical different ways.

After the global financial crisis of 2008 (which was first financial and then economic), many had been talking of a "new normal" that we should have been aiming to achieve, born out of a re-basement of the (mostly) financial dimension of main global markets towards more concrete "real economy," with the emergency of

© The Author(s), under exclusive license to Springer Nature Switzerland AG 2021
C. Scardovi, *Sustainable Cities*, https://doi.org/10.1007/978-3-030-68438-9_1

digital technology as the driving force of wealth creation and geopolitical supremacy. A bit more than 10 years later, well-known strategic consultants are now talking of a "next normal" that will redefine, 10 years later, which kind of next steady state we could settle in—as if life could flow linearly, with some crisis hiccups from time to time. In truth, in our view, a more proper naming to consider for this could be "other normal" or actually "other normals"—if we assume that what lies ahead is much less linear that we think or hope and that sustainability, survival even, is also far from being assured. Because there is a fair chance that many other "evolutionary crisis" will need to be faced by humanity in the near future, including pandemics, climate change, cold wars, and other events, we may not even be thinking of as yet. As stated by Mark Carney (The world after COVID-19 2020), the former Governor of the Central Bank of Canada and of the UK one, the very notion of value will change in the post COVID-19 world and not just for the implosion of stock markets—as there is a possibility that the gulf between what markets value and what people value will close. In the midst of the social and economic disgrace brought about by the pandemic, real opportunities are also revealed by this—in teleworking, for example, and e-health and distance learning.

According to Carney, a progressive (and risky?) acceleration is also taking shape across developed and developing economies: in the switch from moving atoms to shifting bits specifically, as digital and local lives expand and physical and global ones contract. In our other, post pandemic normal(s) ahead of us, everything will change, with resilience (or what Nassim Nicholas Taleb calls "anti-fragility") increasing its relevance over efficiency and with greater attention paid to managing tail risks (those of an existential nature specifically) that typically involve the entire world, are more difficult to predict by traditional risk management, and can be typically addressed only on a multilateral basis. Given such a critical and complex context, it may look a daunting task to discuss how people, companies, and institutions could react and raise this challenge of change. Or it could even appear as a frivolous task the discussion on how cities or companies should evolve to stay cool and relevant and become even stronger and successful. In fact, far from it, these many battles ahead of us—from pandemics to climate change, to geopolitical tensions and cold wars—will need to be fought in the urban areas and city of relevance and by the companies and institutions that have, in the citizens of reference, their anthropological foundation and object of ultimate ontological quest.

1.2 Second Sleep

All civilizations consider themselves invulnerable, writes Robert Harris. History warns us that none is—as the recent pandemic has reminded us all too well. Apparently, all innovation looks irreversible and mostly for the good of humanity, we could add. We may have start feeling that is not always the case. The street towards success, through technological disruptions as most straightforward shortcut—has often shown to be paved with bad intentions and even more awful results.

In Robert Harris' most recent novel, *The Second Sleep* (The second sleep 2019), humanity is waking up again, after an unspecified apocalypse that has hit humankind. It's 1468, and a young priest arrives in a remote village in the British countryside, with the land around that is strewn with ancient artifacts—coins, fragments of glass and plastics (certainly a banal error of the history scholar and author of bestselling novels). In fact, as the thriller unfolds, other artifacts are found, including several strangely shaped rectangular prothesis with a visible screen on one side, the size of a hand, with a half-bitten apple printed on that, that ancients used to consult almost as single source of truth. In a middle-age scenario where humanity seems to have lost event, the faintest memory of the glorious technology it was once able to master, even the observation of electricity, is compared to the work of evil and something that, if pursued, would bring back the advent of the beast and the apocalypse itself. Plastic remains are just odd testimonies of the once upon a time power that men (and women) were able to exert on nature itself—before the un-linear beginning of the end started taking place, in an uncontrollable way. Whether the apocalypse happened because of a global cyberattack (able to block and destroy anything digital), or for all-in nuclear warfare, or for the now more fashionable scenario of total destruction brought about by climate change or by an antibiotic-resistant pandemic virus (coronavirus or else), it is not known in the story—and it does not really matter anyway. The main point of the thriller is about the fragility and inherent unsustainability of the system we have built.

As we still strive to exit the COVID-19 emergency, and the following economic and financial conundrum that will put our countries, developed and underdeveloped, under the huge strains of uncontrolled public debt, unsustainable equity wipe outs, and extended unemployment and poverty, we have ventured, in the previous paragraph and following the hint of Mark Carney, that digital could save us from most of this. We could all become more digital (and local) and less physical (and global) to increase our resilience to this crisis or to the next ones. In fact, we believe this could (luckily) be just a half-truth. Digital and local resilience will certainly become more relevant and shift a bit of lights away from atoms towards bites. However, even as we restart to operate in an unlocked-down environment, we will need to feed things like digital surveillance and e-health check monitoring with the flow of offline information and then feedback intelligence to the physical world, to support its safety and improved resilience. In turn, as the pandemic crisis will stabilize, we could find us in a new world (another normal) where the digital and physical dimensions enmesh and are closely intertwined and—also thanks to that—where the local and global dimension will find a new way of coexisting, synergically and sustainably. This could happen as we could become abler, for example, to reinterpret meta-needs such as "mobility" (once the exclusive realm of categories and industry labels such as "transportation" or "automotive") as a mix of the cyber and physical ones (as we leverage augmented reality and 3D videoconferencing, but also "bullet-trains" travelling in air-vacuumed tunnels) or meta-tools such as "supply chain" (once the focus of global pooling and specialization, lean inventories, and just-in-time production-to-distribution approaches) as a mix of cyber and physical as well (as we adopt 3D printing, with global specialization managed as digital).

The search for another (or other) sustainable normals could then require new, creative ways to mix digital and physical bites and atoms and potentially lead to a different ontological interpretation of most of our business and leisure activities—more based on and driven by a stakeholder-based model of capitalism and development—something Michael Sanders has dubbed as the shift from a market economy to a market society, where everything gest value, in a fair, transparent, and efficient way and hence requiring some kind of market for that (a market for valuing data and information, e.g., as we will later discuss). This ontological challenge, and the very concept of evolving towards a more intertwined and well-balanced development of our cyber-physical dimensions, may appear very vague and ambitious—a good topic for philosophical discussion but not really top priority for policy makers and business managers. In fact, the definition and execution of this challenge goes back to the very down-to-hearth and mundane settings of cities and companies. Whatever other normal we will be able to achieve, it will need to be designed, fought, and realized at the level of the cities—starting from the main ones that are hosting over half of the global population. And, on top of the public institutions' concerns and active promotion of these ideas, everything will likely happen (or not happen) because of the behavior of private companies that are called to redevelop cities, across their digital and physical dimensions, reinterpreting most old categories (e.g., from transportation to mobility, from global supply chains to global-local ones). Not only the relative competitiveness of cities and companies will be determined by this, but also the very sustainability of our social and economic environments will depend on this challenge.

1.3 If Smart Is Not Enough

Cities, as we discussed in our previous work (Finance and real-estate wealth-being 2020), were already becoming the center of our ways to approach, manage, and realize our transformation towards new ways of working and living that were able to maximize our overall wealth and well-being. These was happening, we argued, because of large programs of urban regeneration, "smart city" technological investments, and the planning of unique and attractive propositions for cities and neighborhood that were being rethought around new multiple centralities, with built-in local and global resilience, cyber and physical diffusion. This challenge is still valid, but with a pace of change that need to rapidly accelerate if we want to be able to adapt and survive to the coronavirus pandemic, climate change, the new cold wars, and else. For this, however, being "smartly digital" may just not be enough. Everything is, in fact, becoming digital and driven by data captured when surfing online. Hyper-connectivity and 5G technologies are able to reduce our digital data latency periods to close to zero. And the development of quantum computing offers almost limitless problem crunching capabilities. Also, the evolution of artificial intelligence self and deep learning techniques approaching a GAI (general artificial intelligence) are all indications of the many disruptions to come—with a faster and faster frequency: disjointly from economic cycles, agnostic vis-à-vis geopolitical areas, indifferent with respect to prevailing economic systems. All these disruptions will bring more and more radical dislocations across markets and societies with further polarization between winners and losers, transformed and left behind—ultimately impacting in a very negative way on the sustainability of our civilization. In fact, as it all becomes more and more digital, hyperconnected, super-computable, and intelligence-augmented, it also becomes more and more fragile, prone to devastating disequilibria, and subject to the nonlinear unfolding of terminal events driven by potential facts such as a cyber hack, a human-in-the-loop error, and uncontrolled and malevolent GAI or a simple virus breaking out of the bottle.

All main cities in the world, as Harris reminds us in his novel, are just six meals away from starvation, as all their food chain is fully integrated but geographically dispersed, digitally interconnected and managed with money as a mean of exchange—most of this just digitally based: something that could disappear entirely and become irretrievable and useless, should, for example, electric energy disappear. A simple black out would disrupt the food chain and lead to chaos and unrest, city wars, and migrations back to rural areas. It could disrupt the very fabric of society and harm the civilization developed and matured so far. With it, most of our technology and knowledge of all kind (now mostly digital based, with little remaining printed on paper) would also disappear, forever. In a society where high-tech artifacts are becoming omnipresent and quasi-indispensable, and where individuals have become addicted to the use of their handheld, screen-based prothesis (mobile pads and phones that they consult almost as the single point of truth), everything is becoming "smart." In fact, our way of working and living is becoming so "smartphone" centric (where the smartphone also acts as hub of connectivity and data gathering for many other devices and wearables—from PC to watches, from glasses to bracelets) that we all live, or pretend to live, smart lives. We are all smart persons that work in a smart way and in smart offices, in smart buildings and (of course) in cities that have the aspiration to become "smart cities"—something important and worth paying for. Smart cities are based on digital technologies that fully embrace a new paradigm seeking to capture information in a big data platform (BDP) in a granular way and across their territory, developing analysis on this to provide a number of basic utility services—from mobility to safety, public and private, to residents and tourists, for companies and the administrator of the public good. It all then becomes "smart": smart living and smart mobility, smart safety, and smart public services (reading from the leaflet of one of this smart city programs being pitch by consultants and TMT companies, but also by big digital ones—the likes of Google and Microsoft—not to mention bank real estate developers).

Smart city ambitions are at the heart of large urban regeneration programs and new developments and have a prominent place in the strategic plan of global metropolis like Singapore and Seoul, Toronto and New York, Paris and London, and Berlin and Milan, but also in smaller ones, such as Bergamo or Firenze, to name a few in Italy. Having a "smart city" approach is then becoming a key (or the main?) component of the value proposition of a "battle among cities" that is developing more and more across borders and with the aim to become one of the most attractive places to stay. This search for a total "wealth-being" attractiveness (where the living well proposition is matching the living in a place great for work and business opportunity) is driving most of this, as in an increasingly (digitally and physically) interconnected world, the place that—for its very specific and unique value proposition—becomes the most attractive and sought after will tend to "win it all." The best capital markets city, or sky resort one, the best luxury and fashion metropolis, or senior living one—they will increasingly polarize the capture of attention and, hence, time, interest, and money. This "wealth-being-fullness" will then be reflected in the price of the land and of real estate, and, as real estate as an asset class is making up more than the sum of equities, bonds, and gold at global level, it will reflect in the overall financial wealth of the people and institutions owning this. This financial wealth effect will then attract new investments and drive greater and greater consumption cycles, hence increasing, if not the quality of life, the perceived "good life" of the people living there. A most extended definition of "smart city" could be based on its ubiquitous, high-volume information flows that are continuously happening at the intersection of habitation and computing, and, according to Carlo Ratti (The city of tomorrow 2018), it is dominated by three key elements. The first, "instrumentation," refers to the omnipresent array of sensors measuring environment conditions and movements. The second, "analytics," is connected to its algorithms that use massive amounts of urban data to find patterns and formulate predictions. And, third, it relates to its "actuators," e.g., its digitally controlled devices that can respond to data in real time and impact the physical space—anything from occupancy-sensing architecture to resource-saving utilities, including "air conditioning balloons" that can follow us as we walk in our flat, to warm (or refresh) us instead of the generic space we are living in. However, most "smart city" plans are still based on the digital dimension only and on evolution of basic utilities—such as transportation conditions and energy consumption, safety surveillance, and the access to public services. They involve mostly the digital dimension (in terms of data captured and services offered) and, more importantly, address the "how could the city work better?" question more than "how the citizen could live better in the city and enjoy a greater wellbeing?". They try to optimize the way basic utilities perform, more than supporting targeted lifestyle and addressing physical and spiritual needs that could be interpreted from the analysis of people's "true" behavior—not just the one in the online world but also in the real life, as they work and enjoy life in the city.

1.4 From Smart to Cyber Cities

We all rely on digital technologies and on the ubiquitous connectivity that allows us to spend more and more time online—this trend has become stronger and stronger in the last few decades and will develop further with the advent of "smart cities," made up by smart buildings and smart flats, with smart self-driving cars and a host of other smart-wearable objects—all working as spokes of a smartphone, screen-based hub, to support our quest for working, playing, and living smartly. Alas, the physical dimension (and the spiritual one that is more discernible by analyzing how we behave in real life as well) still matters a lot, whether related to food and entertainment, to our total wellbeing, or to the climate change risk that is impacting our overall ecosystem. Hence, "smart" may not be enough. And the way "smart cities" are currently designed and planned for development may also be just an imperfect, first step towards a more holistic, balanced, and sustainable direction. In fact, in a society where high-tech artifacts are becoming omnipresent and quasi-indispensable, the digital dimension impacts everything, including real estate and infrastructure assets, contributing to make their value more "intangible-based" and "intangibles-driven," defined by the data they can generate and that can be captured, aggregated, and analyzed to look for past correlations and future predictions—eventually, to derive serviceable, timely, relevant intelligence on built stuff, as well as on our bodies and state of mind. But this digital dimension, and the level and sophistication that could be allowed in its interaction with the human species, is hardly captured and expressed by the "smart" adjective. Smart, in the English language means chic, trim, neat and fashionable, trendy, and even snappy—adjectives which are all part of the proposition of smart things but should not stand as its core. In other languages, smart could be even translated in more outspoken (and less positive) ways—described as something that plays well on a trick and mostly to its benefit by outsmarting others ("furbo," e.g., could stand as a good Italian translation—with someone "furbo" that does neither necessarily nor naturally stand for intelligent, hard-working, and team player. It could be someone that makes up with selfish tricks for the lack of the abovementioned qualities).

A more proper adjective to describe how digital technologies are changing cities could then be "cyber"—more directly referring to the dynamic interaction between humans and the digital machines (computers, but also big data platforms—BDP and Internet of Things, IoT, devices), let alone the reference to the fashionable Tesla "cyber truck" promoted by Elon Musk. A "cyber city" would then refer to a place where virtual reality, BDP, and IoT are building parallel "twin" cities based on information and intelligence to provide something more than basic utilities. The different emphasis is not small change. On one side, data can be gathered to augment the level of information that are needed to optimize (say) the traffic conditions, or to monitor safety, or to reduce the use of energy based on occupancy and need. On the other side, they can be developed to offer a kind of easy to use interface connecting the virtual reality and our daily life in the physical realm, with the twin digital-to-physical dimensions feeding back continuously to each other. In a "cyber city," the augmented physical reality gets developed and compenetrated by the digital dimension and vice versa. Data and information are also used to mine intelligence about cause-effect correlations and potentially, based on nonlinear patterns that are discernible for artificial intelligence/machine learning (AI/ ML) applications, to create intelligence that can be used to identify and deliver solutions to hard problems. These problems may also involve what we described as "utilities" but can extend to the many and multifaceted dimensions of the working and social life. A cyber city would then, in our extended definition, cultivate the ambition to deliver solutions across the economic and personal spheres, contributing to the success or failure of businesses, to the development of an urban tribe or social community, and to the overall civilization (or lack of it) of an entire society—hence, it would support the ambition to make the city a winner on a global scale. A recent study by ULI (Urban Land Institute) has provided an interesting framework for summarizing most of the technology trends at play that are promising to revolutionize cities, in Table 1.1.

A "synapsis-based" model (Digital transformation in financial services 2018) made up of what we described in another book as five fundamental pillars would then come to dominate most of the ways of working of the emerging "cyber city." Data and information would be continuously captured in real time and capitalized in central (or dispersed) BDP—and they would be made available to multiple counterparts in a transparent, safe, and economically efficient way for them to work on it, as if they were a gold mine. Traditional and more innovative applied analytical techniques would then become widespread and used by multiple counterparts to extract meaningful intelligence out of this data and to get information with a purpose. On this basis, new model of interactions and intermediation would develop and flourish, connecting multiple stakeholders together—in a more holistically extended and inclusive approach, not driven by the mere arbitrage of information advantages but the quest for solutions. In our example of the "synapsis bank," it is for the bank to evolve from its mere intermediation "buy money—sell money" game towards a more stakeholders-based, many to many one, where multiple stakeholders are brought together to working on a specific hard problem. The value chain is also reconfigured, with the direct, supporting involvement of the many stakeholders and with the final aim to provide a solution that was not in the cards and truly additive to the value of the ecosystem. Solutions, the fourth pillar of our synapsis framework, would then be delivered and sold, eventually adapted and reinterpreted, to the overall community on the basis of trust, the real scarce resource of the digital realm—threatened by cyber risk, fake news, data manipulators, thefts, and so on.

1.5 Beauty, with a Purpose

For all the talk of "smart" things—from smartphones to smart building, smart companies, smart cities, and, of course, smart individual—all coming to dominate our contemporary reality, this term could be interpreted almost as an offensive one, as its "intelligent and chic" interpretation moves towards a "fashionable and snappy one" and then to a "witty and natty" one ("furbo" in Italian—somebody playing dirty tricks to get an unfair advantage on others). We have then proposed to use cyber instead, and with specific reference to the cities of the future. Cyber, and cybernetic (as a dynamic way of humans and machines to interact and change behavior based on continuous feed backs), captures better the idea of a city that is looking to develop better and more inclusive solutions towards sustainable futures where mind and body and digital and physical dimensions are optimized as a seamless dimension

Table 1.1 A new urban technology framework is potentially transforming cities in unprecedented ways (ULI framework)

Digital twins Augmented and virtual realities (and associated city systems models) create digital versions of cities that become part of the city	**Tribalisation** Social media and mobile connectivity create small, fluid tribes defined as much by digital as physical places	**Share and share alike** The subscription and sharing economy extends to ever more of our 'stuff': houses, cars, even clothes	**Smart cities** Tech progress creates denser, more efficient cities, better governance, and the orderly provision of citizen services	**Builders assemble** Onsite construction becomes a final assembly process for large, complex sub-assemblies manufactured in factories	**Smart buildings** The Internet of Things makes buildings and places better machines: more efficient, flexible and sustainable
Digital retail From stores to omni-channel; from things to experiences	**Future of work** Emphasis of work changes from the company to the individual and the project	**Digital manufacturing** Additive manufacturing sees making things return to the city center	**Wellbeing** The monitored and quantified self supports healthy and happy lifestyles and citizen wellbeing	**Mobility-as-a-service** Public, private and active transport become joined-up, pay-as-you-go	**Smart-machine** My house, my car and my robots all understand me

AlixPartners

(or driver) of happiness and meaningful pursuits. Cybernetics is in fact referring to an interdisciplinary, mathematical-based approach that is aiming to study the complex systems of living beings, whether of natural or artificial origin. Hence, cybernetics cities, or cyber cities (in short), can better express the complexity of the "human + machine" coexistence and its most immediate fall out, that is, the convergence and relationship between physical and digital.

Where "smart cities" refer to multiple applications, mostly related to everyday utilities which are relevant but not crucial in the way we could design and perform our new ways of living, "cyber cities" come to support more productive and enjoyable lives of its citizens: they support the pursuit of a meaningful, total (mind and body) "well-being with a purpose." Cyber cities are then supporting a more human-centric approach, where citizens as producers and consumers (and investors as well) are called to play a crucial role as co-designer and actuators of urban developments and regenerations. Cities are, according to Carlo Ratti, by definition, plural, public, and productive (The city of tomorrow 2018), and they are created (or evolved) by society itself, with design by mutation happening as a collective process. Cities' evolution happens as driven by the energy of the crowd and relying more and more on the intertwined forces of the digital and physical realms. It's a mix of bits and atoms, according to Ratti, that needs to be reconciled at city level. And, according to us, it's on this reconciliation that greater resilience and sustainability can be built and ensured, to face off crisis like the pandemic or the climate change ones. This is a huge task on which public Institutions and private sectors need to work synergically and where multiple industrial sectors and players can have a role. In fact, we have argued in other works (Finance, real estate and wealth-being, Claudio Scardovi 2020) how finance and real estate, along with technology, can become the acting forces of the redefinition, regeneration, and evolution of large metropolitan cities, as plural, public, productive, and enjoyable—hence highly attractive and sustaining the creation of a shared wealth and well-being. But it is this "well-being with a purpose" that should ultimately come to define them.

It is the search for this ultimate city's purpose, along with its beauty, that drives its successful urban developments or redevelopments and the many regeneration programs of aspiring "cyber cities." The "smart city," as domain of digitally integrated urban spaces, with ubiquitous tech suffusing every dimension of the physical realm, should develop with an active participation of its citizen that are contributing to its co-designing and will end up co-living within the city, working in this and spending leisure time in it—they should then demonstrate (and be guided, by professional people) a strong interest in its aesthetic dimension. Beauty is, in fact, critical for the marketability of a place, and the first indicator of its place-location (or value proposition) strategy. Only beauty can support the full realization and sustainability of the wealth-well-being-fullness of the city's citizens. Only beauty can express (in a synthetic yet very powerful way) the finally recovered mind and body equilibrium of people that is a key component of sustainability. Purpose and beauty should therefore suffuse and influence every dimension of the cyber city, along with ubiquitous technology and its ecosystem of sensors collecting information from the urban space. Purpose and beauty should inspire the design and management of open (big) data platforms, both centralized and decentralized, as a basis for the formal and informal collaboration between citizens and governments, where machines and crowds and companies and institutions work together to innovate. Data-driven feedback loops among these parties can in fact be used to design and actuate several test beds of innovation, across the many dimensions of the wealth and well-being-ness of individuals. Purpose and beauty should then drive the actions of the network-enabled actuators that, by synapsis, can transform the urban space and allow a real, meaningful, incremental impact on the way we experience our work and we enjoy leisure time and socialize across tribal (digital and physical) communities. Purpose and beauty could provide the special passion and those ever-changing emotions that can make a large city, or even a small place, unique and in a world of its own. Purpose and beauties would therefore underpin the increased resilience and sustainability of cities that are at stake today and key goals to achieve.

1.6 Cyber Cities Are Not for Cyborgs

The design of cyber cities, the design of their architecture and of their built environment, and their technological infrastructure are driving an always-on, real-time, fully connected living, in an enmeshed, sentient, and seamless cyber-physical space. In this, bits and atoms become suffused, and multiple machine-to-crowd-to-individual communication flows can develop, across multiple interfaces. As screen-based (via PC, smartphone, watches, and so), or as touch-sight or voice activated or as directly implanted in the people's body and brain, with intertwined biology and technology to create an almost new human-machine system, with Elon Musk obviously working on a cyber-brain as well. With his Neuralink project and related company, Musk is aiming to implant flexible "threads" into the human brain to allow the transfer of big volumes of data. This is not to create cyborgs, or to allow AI to take over human brains, but to allow the two to coexist in symbiosis—again, as a fusion of the cyber and physical dimensions of an extended, augmented, and suffused reality. It may look still look as sci-fi as of today, with a few years ahead before the first, actual implant. But in truth, the relevance of screen-based digital devices is already diminishing, and they could progressively and potentially disappear in a digitalized-physicalized world. IoT of a different nature are instead already taking more relevance, with new devices doing the ubiquitous computing trough the digital cloud to which they are connected, with batteries always on, that can be easily recharged by radio waves and become almost limitless in the duration and useful life.

As IoT ubiquitous devices shrink in size and with batteries always on and charged, their cyber-physical dimension gets embedded everywhere, including inside our body and brain, with intuitive to use, not invasive to implant, almost invisible, unobtrusive new devices—such as eye lens or glasses or listening and amplifying devices that can work as a computer for us, with a navigator actioned by our sight, hence able to provide, for example, the full screening, background checking, and any other relevant information in real time on the people standing in front of us, not to mention the real-time translation of the foreign tongue in use. Not only that. At the individual level, new human-machine integrated systems (from exoskeletons that can allow paralyzed people to walk again, and to other people to run faster) to brain implants (as foreseen and invested by Elon Musk—with first trials soon undergoing) that are connecting IoT directly with our synapsis, all of this will allow us to reconsider our human mind and body relationship and coexistence, potentially removing some of the biological and physical related constraints we thought were just unsurmountable (we also thought that men could not fly on thin air or spend time in the sea depths). As it's the case for cities, also the possibilities of evolution for people are almost limitless and more than just "smart." We can become smart-digital people if we leverage a few tricks from our smartphone, as proxy of a prothesis we can take with us when walking or doing activities for work or leisure. But a cybernetic human is something more, as he can evolve via the balanced extension of our cyber and physical capabilities, extending our many powers and augmenting our will and power almost at whim: from our computational skills to our memory and ability to translate out thought in infinite languages and from our strength and speed to our ability to see and hear at far away and microcosmic levels. We are born human, but we can become trans-human if we can leverage technology at best, in a cyber-physical way, extending our "persona" without necessarily becoming an AI-controlled cyborg.

Based on these augmented capabilities and almost limitless powers, we can in fact re-imagine our identity, redesigning our extended and more functional economic and social functions—at individual and group level. We can leverage transplants and implants to reinvent ourselves with the aesthetics dimension (as in the case of cosmetic surgery) already preceding the redesign of other, more functional, and less frivolous assets of ours. As cybernetic individuals, living in cybernetic cities that can enable in full our augmented cyber-physical capabilities and almost limitless powers—cyborg-like but not as cyborgs—we can end up living a new condition and paradigm of human existence where physical and digital suffuse and biological and technological dimensions get seamlessly intertwined, powered by applications sitting over the cloud. How would such a setup have been impacted on the recent pandemics if already in place? How could our behavior be influenced to prevent climate change (as an inherent anthropological phenomenon) if digital and physical were always interconnected? How could we all work and live social and economic lives at an order of magnitude more productive and information rich, if we all had already Elon's threads implanted and well-functioning in our brain? With the new technology, not necessarily screen-based, we can project and multiply our self to a conceivably infinite degrees, internally and externally, becoming meta-human, with tech inside and prosthetics outside, cyber and physical all over the place, all inextricably enmeshed and coevolutionary, and living in cyber cities. However, cyber cities are not for cyborgs. And the IoT and machine-based developments are not suggesting a prevalence of the robot, even if the risk of a GAI (general artificial intelligence) super power that takes over the humanity do exist. For the sake of this discussion, we will just argue that "smart cities" need to naturally evolve into "cyber cities," so that cyber individuals can develop into meta-humans, with meta cities in turn emerging to reconfirm the centricity of humankind in the context of the contemporary civilizations.

1.7 Death of the Smartphone

With the advent of mainframes, of the world wide web, of the digital cloud and with the progressive maturity reached by AI and quantum computing, it has become obvious, on one side, how the physical realm can no longer be considered prevailing, let alone absolute. On the other side, the digital one, the object of many "hyped" transformations, is also not enough to ensure the pursuit of economic and social success and the development of super competitive ecosystems. In the first days of the "lockdown" period, we may have come to believe (or just hope) that more digital connectivity and more computing power could have done the trick, to allow us to live the same life, as productive and enjoyable, just more online and less offline. We have then come to count the days towards unlocking, to take back our outside-our-homes, physical dimension—then realizing how much our digital twin has become so critical to "just get back to the old normal." We need digital tracing, for health safety; continuous monitoring, for enforcing social distancing; and so on. But also, we need our cyber twin to keep the extended mobility, communication, training, dating, etc. that we have come to learn and appreciate when locked at home. Most of this can now be offered and delivered via our smartphone—our prostatic, digital extension and everyday companion. For sure, we own globally more than 5 Bn mobile phones and counting, and we spend a good 3–4 h per day staring at a screen the size of our hand. Certainly, we also feel subjugated and sometimes frustrated by our digital and "smart" prothesis, as it becomes so encompassing to allow us to communicate or pay, consult information or manage our domotics-based home, read a book, and play music as well as drive, run and exercise, and meet and date people for work or leisure. Sometimes, we become so overloaded by applications and bombarded by information that we seek some "data diet," to recover in full some of the physical sensations that have been acting for us, since time immemorial, as our interactive interface with the outside physical environment.

But, as we described, a new generation of IoT devices, all supported by a more and more ubiquitous connectivity and by powerful applications available over the cloud, will greatly reduce the "power of the screen." It will potentially lead to what we have been dramatically dubbing "the death of the smartphone." In this IoT-enhanced, nonscreen-based environment, we can recover and augment, at the same time, part of our set of physical sensations. This is not a dream of a distant future. Take, for example, Mojo Vision, a startup from the Silicon Valley that is developing augmented reality contact lenses that will give users real-time information from the web. It may have as end game the creation of an interactive world that can be layered on top of the real one. Mojo Vision's contact lens includes a micro-LED screen, a microprocessor, wireless communication, and sensors that could be connected over the cloud. Its lenses have then a digital display that is navigated with eye movements (Financial Times 2020). Users glance at functions on a "home screen" to select them. They can then be activated by staring at them for a few seconds (something like a double click, while a side glance is like dismissing the information or function). Potential use cases are almost limitless and intuitively so. They could, for example, provide subtitles for conversations in foreign languages and fitness statistics for runners and integrate vision with voice commands. They could be used for cheating at exams or to look super-smart when you really aren't, but could also work as safety devices, alerting when driving, cycling, or skiing if your lens (and the AI behind them) are spotting a fair chance of an upcoming crash. They could also help in recognizing faces and pulling relevant information on individual (with suggestions on what and how to say to "double click" and get a "like" on her relationship) and support policemen in recognizing dangerous criminals (this is just the case in China, albeit through glasses). They could also support workers on their most critical tasks—think of a surgeon doing open heart stuff. Even image sensors that could be incorporated in extension these lens to allow, for example, a 360 degrees fish-eye perception, with all the computing power hosted on the cloud.

On a similar token, we could rediscover our flower-smelling capabilities, with the help of appropriately augmented "digital noses" (already existing and able to sniff explosives hidden in bags at airports or drugs concealed under the hearth and even the quality of a certain truffle). Similarly, we could hear, with the help of "digital ears," the sound of waves that are coming from a seaside a hundred kilometers afar from us (and with no "white noise" in between, because of our digital hyper-hearing capabilities) and so on. We could then also recover and multiply all of our physical forces, with their powers multiplied by the embedded technology—IoT implanted inside, with networks and applications running outside and all pointing to the brain and (quantum?) computing capacity sitting over the cloud. All these high-tech components would provide solutions mixing digital and physical data in real time—extending us to meta-humans, not substituting us with cyborgs. More importantly, in this co-development between context and content (if we can sniff better, context, and recognize immediately which flower is that, content, if we can hear better and see from much far away, context, and understand what is being said, even if in a foreign tongue and associate a detailed identity to a face, content), then a new info-telligence can be created, putting into correlation multiple contexts and contents to derive then serviceable truths.

1.8 From Cyber to Meta Cities

This info-telligence (as a mix of new, relevant information created by basic data and superior intelligence) is going to dominate, in our view, the development of cybernetic cities and their evolution into the winning meta cities of the future. Meta is, by definition, something that is occurring later and in succession, being situated beyond something that it's trying to surpass and englobe by extension, through a more highly organized design that is comprehensive (covering both cyber and physical dimensions) and transcending (extending the values of the two in a nonlinear and highly accretive way). Meta cities would come to collect huge amount of data from both the online and offline world, using this trove of data to better understand how people behave and interact with each other, at work and during their free time, in the digital space as well as in the physical one, all granularly mapped and precisely overlaid, forming cyber-physical communities that could "swarm" online and meet offline, or in a number of alternative mixed systems sitting in between. Suffusion technology can in fact allow these mixed cyber-physical systems to sense, capture, recreate, and then manage and change the real world we live in, extending it beyond the physical constraints of the lockdown or augmenting it beyond the lack of data we would otherwise face when locked out of our home or office. Meta would give a practical representation of this outcome, through the tribes and communities of people that populate them. In our vision, a meta city is one that merges cyber networks and physical spaces at best—and where cyber and physical configurations actively influence each other, with no leadership of the one on the other and with multiple space—time relations available for extended manipulations and interpretations.

In fact, distance is not dying (as famously argued by Frances Cairncross of *The Economist*) as we are still dependent upon being present in a given location and at a given time. But time and space are taking up a further, holistically augmented meta dimension, where quantum experiences (running in parallel on cyber and physical dimensions) can be captured, analyzed, and deciphered by quantum computing. On this ability to play both cyber and physical, international cities will fight their competitive battle, with meta cities becoming bigger and bigger, polarizing and attracting most of the global population and creating more wealth and well-being—in a highly polarized (between cities) but hopefully more inclusive (within the single city) way. In meta cities, the individual meta citizens will show a more explicit awareness of themselves, as member of a certain community—and of an extended ecosystem that is becoming more and more digitally and physically intertwined and single "fusion." They will then act as human magnets because of them becoming cyber and "much more than smart," but also because of their understanding of the importance of physical interaction among people and between these and the environment, mediated by powerful BDP that will be run by almost omniscient algorithms able to support continuous mutations and evolutions by design.

In meta cities, meta companies (or cyber and physical—cy-phy—companies) will also play a central role, with their way of being interested not only on how things are, but also with how things could or should be. More than companies active in the transportation or health or energy sector, for example, these meta companies should be focused on delivering solutions to address people's -"mobility," "well-being," and "empowering" needs. Our mobility in the city and elsewhere, cutting across our professional and personal dimensions, could be addressed by a mix of communication and travelling tools, leveraging augmented reality and real-time, 5G-powered IoT devices. Digital communication could then be backed by physical delivery ensured by drones or 3D printing (e.g., if we work or design from home). Our health in the city and elsewhere, cutting across our psycho-physical productivity at work and home, could also be addressed by a mix of basic body functions that are continuously monitored and tested via IoT (e.g., blood pressure and testing—could be done by eye lens or bracelet, with digital toilettes already checking on our organic remains) and extended to our training activity (when doing "spinning" with crowds, with our static bike and PC in our bedroom, or pedaling on the mountains, with a chat-bot trainer always online to provide guidance and direction). Finally, our energy in the city and elsewhere, that we need to keep warm or cool off, to move around or do computational activity, shift weights or run faster, it could also be reinterpreted and addresses as our need, across our professional and personal dimension, to have greater power than our psycho-physical constraints could warrant. This "empowerment" could be addressed in much more personalized and effective ways, with "intelligent" shirts that keep us at the desired body temperature and level of humidity by monitoring these and reacting accordingly, but also being able to deliver energy that is recharged via radio-frequency wireless charging technology waves or else and with physical power, to weight or move, delivered by e-bikes, taxi-drones, or e-exoskeletons (these, also already available, even for running across the Alps, for less fit people).

In short, in meta cities, meta companies could pursue critical needs and relevant solutions by reinterpreting, in a most extensive ways, the meta-industries that are required to address and deliver them. "Mobility," as a need, can be better fulfilled as a mix of products and services coming from (traditional, likely passé) industrial sectors like transportation, automotive, aerospace, telecom, media, and so on. "Well-being," as a need, can be better fulfilled as a mix of products and services also coming from (traditional, likely passé) industrial sectors like healthcare, pharma and biotech, hospitality, leisure (for gyms, whether online or offline, for the likes of Peloton or Zwift). And "empowerment," as a need, can be better fulfilled as a mix of products and services also coming from (traditional, likely passé) industrial sectors like oil and gas, real estate, fashion, and again transportation or heavy equipment. We could easily end up reinterpreting and targeting all critical needs through a blend of mixed offerings cutting across multiple industries. The winning meta companies that would be better able to deliver the best "mobility" or "well-being" or "empowerment" proposition in the meta city for us, and at a very personalized level, would then likely act across the blurring boundaries of several industry, but also being characterized by a common denominator, e.g., their ability to work with data, both from the offline and online worlds, to harness the best information available and extract superior intelligence to produce "serviceable truth," something objective enough to be put into action to get some valuable outcome.

More than that, winning, meta companies could also become very involved and master-of-design and aesthetics—with beauty becoming a key component of their widespread adoption and commercial success (how much is Apple, or Dyson, success determined by their cunning design? Their cool feeling? The inherent beauty that links a superior digital technology with the physical context—the city first and foremost—in which they are destined to be used?). Augmented glasses won't be successful until they became beauty and fashionable to wear (see the latest Bose—

glasses cum stereo that are top fashion and sell at the same price of top fashion glasses—with technology as "on top" and almost for free). And exoskeletons will be a thing of the fantasy, until we won't feel like stupid when wearing it but super-cool as Schwarzenegger when playing Terminator. This is not a small point, as aesthetics and beauty, of anything commercial that is proposed to people, will need to be judged within the context of the meta city also being planned and designed. Elon Musk Cyber-truck (the electronic pick up envisioned like a Star Trek recreational vehicle) may look like hot stuff but try and ride this in downtown Milan, with its 5.9 m length, not to mention its almost 2 m width. For cities in order to be successful, beautiful architecture and a nice overall setting have always been critical. This will also be true, in the context of meta cities, with urban regeneration happening in parallel with the digital embedding of the surroundings. Beauty, we could say, will save the world, but a little bit of help from digital innovation, suggesting new ways of living in equilibrium with the environment and in more socially sustainable ways—and with a sense of purpose suffusing and inspiring the convergence of the cyber and physical dimensions. For example, "cy-phy" mobility will not be just about making traffic lighter and less polluting or saving time and reducing accidents—it will be inspired by other, more ambitious pursuit such as "making travel an entertaining experience" or "a maieutic, soul searching journey." More ontological questions could then be addressed, on top of down-to-earth issues and other very practical hard problems. These ontological questions would form the basis, in the meta cities of the future, of the search for truths that are defining, challenging, and offering an higher sense of purpose, to the human condition and being. With suffused technologies, existing all around and within us, that would allow us to recover and augment our senses, capture data and info-telligence from the online and offline worlds and live with our cy-phy community of choice as we feel like and in pursuit of our vision of well-being-fullness.

1.9 A "Cy-Phy" Humanism

Will the coronavirus pandemic, or climate change, or the next existential challenge to humanity to come kill the attraction of megacities? This is one of the emerging sentiments discussed by media in the aftermath of the COVID-19 outburst (The pandemic is killing the attraction of megacities, Camilla Cavendish 2020). In fact, big urban centers are emerging as the new plague pits—from Milan to Madrid, from London to New York and even super "smart" Singapore and Seoul have all been hit severely and dramatically by the health crisis. People density if, of course, one of the main allies of the plague. And its current main enemy, lockdown, is better practiced when you are in the countryside, with larger flats you can afford to rent or buy, for the same money, likely with a big garden with them and large public space where you have a fair chance to meet a cow, or a leper, but not many other human folks (animals are mostly immune to COVID-19). If social distancing does not allow you to work at your company's prestigious office (in the words of Jes Staley, CEO of

Barclays, the days when it looked like a good idea to put 7.000 people all working together in a skyscraper in Canary Wharf are maybe gone), then what is the point of commuting every day to the big city? And if social distancing is killing the scene of main street, with empty (or fully closed) shops and restaurants, what is the point of putting up with pollution, traffic, crime, and a much higher cost of living (not to mention the cost of real estate, likely to be 10- or 20-fold the one in rural areas)? Are the polarizing effect of cities and secular urbanization trend still worth discussing?

Almost overnight, writes Camilla Cavendish, cities have gone from being places of dreams and ambitions to fearful symbols of mortality, with the truly rich retreating to the countryside or leading a self-imposed quarantine on a yacht in the Caribbean. Are they going to bounce incrementally back when the restart phase will be in full swing? And we will reach an "other normality" of work and life? Certain economics would suggest not. For example, manufacturing has, since time immemorial, migrated in the suburban area. And what is the point of paying high rents in the financial district, if services are better dealt with from home via smart-working (another "smart" thing, almost inevitably). In truth, this was already possible before the crisis, with the PC and the connectivity already available. However, not only the pandemic has brutally accelerated our switch from phone calls to video calls and from in-person meetings to virtual ones because of the increasingly available "smart" technology. But also, a radical cultural shift has taken hold of our ways of socializing. It's OK to see you working from your bedroom, with your cat strolling over the desk and a children screaming in the background. If you are working on your report from home, it's not "lucky you" anymore, and no "half a day off today?" would be acceptable right now, not even as a joke. In fact, we have all come to understand that, working from home, we are likely much more productive and working harder and longer and we may be even getting more creative at that. We are not wasting 1 h and a half of commuting every day to check emails from our shoe box in the office. We are not contributing to CO_2 emission because of this. We are not taking flights to other megacities simply to shake hands (prohibited now, anyway) of senior clients which have become so detached from content as we are (because of the time we waste in airports and taxing across trafficked cities). Not only that—as, along with increased productivity, we may have more time to spend with the family or to take a jog in the countryside (without the risk or running against a crowd of white collars also exercising at Hyde or Central Park). In short, we could have more time and space for ourselves—eventually, the globe is so huge and vast that there should be no reasons to have all of us cramped in so little space. De-urbanizing, it seems, could then be good for our wealth and well-being.

If the pandemic, and the other existential challenges to come, are all calling into questions the value of city life (and of its real estate stock), not all cities will be necessarily required to suffer in the same way. Eventually, we believe that some of them will actually be able to prosper and become more attractive and even more polarizing, pursuing wealth and well-being, the health of its citizen and that of their environment. But in order to that, they will need to radically change—not just be "smart." The declining cost of distance (a fact, given evolution in transportation) has been challenged severely by the monetary and opportunity cost of not being distant (if we live in the city, we pay higher rents, we breathe more polluted air, we spend more time in traffic and with an higher cost of living, etc.). This trade-off, after the pandemic, is quickly switching in favor of urban areas and playing against megacities. Becoming "smart," by offering a few digital-based utilities, can help a bit, but will not address the crucial question of why people should keep living in a second floor 50 square meter apartment in Knightsbridge instead of a 300 square meter detached house with a big garden in the Yorkshire region. A radical rethinking of megacities is now overdue, if they want not just being competitive (among each other) but also survive the potential "ruralization" to come. For this, being digital and "smart" is not enough. Certainly, technology will be of a great help, with the development of IoT, 5G (fifth-generation mobile networks), and AI. Leveraging all of them, "smart" cities will help in cutting traffic congestion, improve public safety, and protect the environment. They will monitor sewers, air quality, and rubbish, using smart streetlamps to save energy based on pedestrian volumes, with sensors embedded on roads and bridges to track vibration and humidity for early warnings on potential damage (something notably absent in the Genoa's Morandi bridge tragedy). For all this, technology and IoT specifically will help a lot, but won't be the end of it.

From a current global IoT network of 14.2 Bn devices (2019, according to Gartner), we will get to 41.6 Bn by 2025 (according to IDC) (Internet of Things: smart cities pick up the pace 2020). Still, with a 5 IoT on average per head, this target looks hugely expandable, as it does the progressive shift of the focus from infrastructure building and application development in a localized-company level fashion, towards the migration over the cloud of most of it, to progressively reach global scale and excellence, on data but also on the use of AI techniques. The development of the cloud itself, specifically, is also a critical driving force of this smart-to-cyber evolution to recreate an augmented reality where digital and physical are neatly overlaid and in close coordination and that can be actively managed to lead to solutions design and delivery—as neatly summarized by Thomas Kurian, the CEO of Google Cloud as he describes in a YouTube interview the three phases of infrastructure building, lift and shift migration of computing, and then solution design and delivery as the path forward that his company and the industry are pursuing to ultimately transform everything. This solution design, largely happening on the cloud, will drive change in a dynamic and interactive way, forcing smart cities to become cyber (or better, "cybernetic") and likely developing infrastructure solutions on the edge, localized in the city itself and not necessarily centralized and run by the municipality (or by the owner of the real estate assets and of the land). Cloud solutions will then support the pursuit of grander visions and larger ends—what we described as the ontological dimension and larger than life pursuits addressed by meta cities—with the cloud acting as the very intangible, almost metaphysical, single point of truth, or of superior and serviceable truths, in more and more quantum-like world. All of this requires a much more ambitious approach to smart city development and what we dubbed "cyber" and "meta."

By 2023, Gartner (The future of connected cities 2020) predicts that 30% of smart city projects will be discontinued and for three main reasons: for the inability of the technology to cope with growing demand; for the residents' worries about how data is used; and finally for the value delivered by smart cities being less than expected. We can now add that a major pandemic, some radical climate change, and even a simple cyberattack or "hack" could also be the reason for the premature death of smart cities (and not only)—as fictionally suggested by Robert Harris' future-backward novel. To avoid this fate, megacities need to develop fast from smart into cyber and meta ones and put people, and their physical dimension, back at center stage. People are the crucial actors of the urban development we need to carefully design and execute to overcome existential challenges as the COVID-19 one and fight the potential "ruralization" trend to come. It follows that data-driven models should be also be applied not just for the optimal functioning of utilities, such as energy, transportation, and safety but also for the support of higher and more meaningful pursuits, including holistic mobility, health and well-being, extended empowerment, expanded learning, multidimensional discovery, and so on towards the full realization of our-self, to reach in conscience full a body and soul well-being-fullness. In this human-centric approach, data and the info-telligence that can be derived by AI can support creative and aesthetic design and help set up easy mechanisms for crowdsourcing the future of the city—based on mutations and selections which are ultimately decided by people themselves. A new cyber-physical humanism is at the basis of all of this, with technology allowing people a deeper and better control of space and time and for their greater engagement and empowerment as citizens. Ultimately, big data and AI should be able to harness the world's collective intelligence to solve its most intractable and relevant issues through cities. A new model of info-telligence economy could then flourish and prosper and support the advent of winning cyber-physical companies and individuals that will be able to lead on the disruptions and unfathomable value creations to come.

References

The world after COVID-19, Mark Carney, The Economist, April 20, 2020

The second sleep, Robert Harris, Hutchinson London, 2019

Finance and real-estate wealth-being, Bocconi University Press, 2020

The city of tomorrow, Carlo Ratti and Matthew Claudel, Yale University Press, 2018

Digital transformation in financial services, Claudio Scardovi, Springer 2018

Finance, real estate and wealth-being, Claudio Scardovi, BPU 2020

Financial Times, Tech start-up develops augmented reality contact lenses, Patrick Mcgee, January 16, 2020.

The pandemic is killing the attraction of megacities, Camilla Cavendish, Financial Times, May 18, 2020.

Internet of Things: smart cities pick up the pace, Nick Huber, Financial Times, January 29, 2020.

The future of connected cities, *Financial Times* special report, January 29, 2020

Chapter 2
A Cy-Phy and Fairer Economy

2.1 Digital Is Passé, Without Physical

We all think of us as "digital," and, of course, we all feel we are pretty "smart." Indeed, the two attributes seem to be closely interlinked, at least in our imagination. You cannot be smart in the business and social worlds, if you are not mastering digital phones and watches, home pods, and pads and connecting your car and bracelet via intelligent apps or your glasses even. We are all living more and more digital lives to augment our physical senses, leverage our resources, and stretch our productivity—we end up almost believing how digital could make us immortal and able to reproduce our thinking and ways of communicating to our dears even after our physical body will be gone. Indeed, digital immortality is pitched as a commercial proposition by many a Silicon Valley's start-up (Adam Gabbatt 2020) Eternime one among many others (Dave Schilling 2016). Not only us as individuals but also, almost inevitably, our company may have launched some sort of digital transformation "navigation program," as part of a digital transformation that is encompassing its industrial sector of reference and city of location. Even entire countries and their policy makers are now pitching themselves as "digitally driven" and "digitally minded," when asking for taxes or votes, as they promise to bring their citizens in the digital age—assuming they are still analogic today and doing complicated math with pen and paper. Here is a bold prediction, not based on the analysis of big data and not coming from an AI app: "digital," as a key word, noun, or adjective, will fall out of fashion soon and for a few good reasons.

First of all, "digital" and "digital transformation" are also part of a typical "buzzword lifecycle"—as most people have come to use them most of the time, almost continuously and usually not really having a clue on what they are all about or stretching their definition at pleasure to accommodate their objectives of the day. Hence, we would expect (and maybe hope) that the "d" and "d+t" words will fast become almost unbearable, meaningless, and rapidly "passé." By the way, "transformative" is often used to infer some "transformism" or "transformant behavior"—

© The Author(s), under exclusive license to Springer Nature Switzerland AG 2021
C. Scardovi, *Sustainable Cities*, https://doi.org/10.1007/978-3-030-68438-9_2

which may have a negative meaning, particularly in certain languages, when it refers to people that want absolutely to change everything, because they ultimately don't want to change anything at all. Aside from the "buzzword cycle," there are other more rational reasons as well to start dumping the terms. First, it's not that companies have been analogic in the last 10 or 20 years after all. Second, even us, as consumers and investors, as white collar or blue collar workers, have been starting using digital computing capabilities many years ago, with the advent of the hand calculator and personal computer (digital refers to "0"–"1", the only possible, alternative, and unique states that data scientists use when coding information and that have been at the basis of the modern computing power developments, smartphone included). Actually, new disruptive technologies are now in the making in the computational field—namely, quantum computing, which is based on the infinite possibilities offered by quantum physics to describe a state of a person, object, or information. They will truly make this "digital" term, based on "0 and 1" out of fashion and passé.

More importantly, "living digital lives" or promoting "digital behaviors" and digitally "transforming companies" or even entire countries are just a set of very partial statements—not even recommendable per se, as they all leave out of the equation the other, critical, relevant, and ontologically important physical dimension. In fact, no matter what the "digitally virtual immortality" pitched by Eternime and others could be or ever become, our life in the real world, as we live and work in flesh and bones (and with brick and mortars for companies), is still very relevant and not substitutable by bites. Similarly, we have argued in the previous chapter, we live and work in cities that, no matter how smart or cyber they could become thanks to technology, are still made for physical people and for physical living as well, with very relevant real estate and public spaces and where the surrounding quality of the environment, climate included, still matters a lot, for our wealth and well-being. This is not to say that digital (here used in its enlarged meaning as "anything related to new technologies based on big data, AI based applied analytics and leveraging virtual, extended, network-based connectivity") is an irrelevant dimension to consider for our own evolution. In fact, our hypothesis is that digital, along with physical, are the two key dimensions to consider, complementing each other and both supporting our pursuits of relevant solutions and contributing to the fulfillment of our every day needs and ontological ends. They are made by bites and atoms as well, immaterial but still material, and looking for a total wealth and well-being that should include not only information and intelligence but also physical and biological dimensions.

2.2 Close Twins

While we wait for the maturity of the quantum computing technologies, we can discuss how digital and physical dimensions can developed and be managed to work closely together or, even better, work almost as close twins. Because, for a start, for each of our physical self, there is a digital twin that is just growing and developing in full, with the development of the cyber city. Everything, and everyone, can in fact be replicated as an alter ego in the digital world. From toothbrush to heavy equipment, to companies and even individuals, examples are limitless. They could even include critical body parts that are replicated as digital twins, to be used, for example, as virtual copies of the patients' hearts. These hearts' digital twins can be used by doctors to diagnose and treat cardiac disease or test new pharmaceutical products. Building a digital twin of a person's heart would first require him to do a variety of sensors (The heart's digital twin 2020). Then the data captured from these would be turned by software in a computer simulation with detailed information on how the organ is working and the way the blood is flowing within it and how this is changing in response to certain medical treatments—all kind of simulations could then be made possible and with little invasiveness in the process. Offering a continuous monitoring of patients and with preemptive simulations available, these digital twin hearts can be used to test different trainings or treatments, diets ,and behaviors—they could be tested when the patient is exercising or eating and working or sleeping, with AI techniques employed to predict patterns and spot rare diseases that come usually undetected by traditional medical analysis. This is for the heart or any other "normal" body part. But what about the brain?

Given a sufficient amount of data collected or collectable in the future, and the required connectivity and computing power, everything and everyone could be match-made and with a digital twin brain (now digital, eventually quantum). This digital twin brain could then be analyzed and tested, to find the best ways of training it, to make it more powerful and cunning or influenceable, and to make it weaker and more subject to manipulation. The potential consequences on the way we live, work, and evolve as humans would be just mind blogging. For a start, digital (or, better, cyber) twins are much more than mere replicas, as they are linked via a multitude of sensors and wireless connectivity to their real, physical twins, producing oceans of data that can be captured and analyzed on a real-time basis. These information can then be used not just to understand why things happen in a certain way or to monitor and now-cast events of relevance (e.g., the spreading of a pandemic virus in the city). In fact, they can also be used to predict the future or influence the present (or both): anything from consumer behavior to work patterns and from political elections to geopolitical warfare. Not only the notorious cases of political influence and manipulation that have been recorded, based on the mining of social data (e.g., Facebook and Cambridge Analytica), may have led to the rise of populism. But also, the real-time use of information and AI analytics has already been used and brought to bear in the context of wars fought by unmanned drones and able to hit a (still human) target with almost surgical precision (e.g., the killing of the Iranian head of militaries by a US drone in 2019).

These are just some of the big events that make headlines on the media. In fact, many other "use cases" could be considered, designed, and run, at the microlevel of the individual and of the company, in cyber cities that are born as hyperconnected (or rapidly reconverted as such) and that can be managed through big data and by AI apps. All cars and other vehicles can be mapped, monitored, and managed based on their cyber twins, to run optimized mobility, with limited traffic, increased safety, and a lower level of oil consumption and hence reduced pollution—even better, of course, if driver-less cars are going around and can be directly controlled and optimized by some central brain that is pooling, mining, and leveraging big data over the cloud and back to the street. The same optimization can happen if we have, as pedestrians, via our smartphone, smart glasses, or smart watch (or via an implanted chip) a cyber twin at our disposal that follows and perfectly replicates us on the other side of the screen, informing and directing our most productive behaviors in turn. AI-based mobility apps, working on built—public or private—big data platforms could then help in dynamically changing the two-way/one-way flow of trafficked public areas and even the sidewalks' width as shown by dynamic lights and colors. Not to mention residential, mixed use, and shopping centers and other commercial spaces—down to the level of the small merchant shop. Rents and asking prices for residential properties could be dynamically adjusted on the basis of the perceived and qualified attractiveness of the place. With more people spending time there, in a more enjoyable way and with an estimated, higher purchasing power, properties should become more valuable—given their other technical characteristics, the competition of comparable assets, and other variables influencing the real estate demand (e.g., lending supply and available income).

On the consumer and commercial side, all kind of marketing levers—product, pricing, promotion, placement, etc.—could also be dynamically adjusted, in real time, to take into account the offline customer behavior and other unstructured information to optimize the whole commercial "physical" and "offline" funnel. If more and more people are walking in a supermarket, maybe discounted offer should be taken off. But if they are just crowing around a specific shelf, maybe pricing should be dynamically adjusted to make sure their interest and willingness to buy something (and the management of the overall lay out of the shop) are optimized. And if it's hot and humid like hell, drinks should be shown first (and kept refrigerated), and hot chocolate should not be actively promoted as a killer product. Inventories could then be adjusted and managed to be kept to a minimum, with a just-in-time (info and AI based) approach (but taking into consideration the risk of extreme events like pandemics, also monitored through big data and AI). Even our working environment would get mirrored, with perfectly replicated colleagues monitored to optimize smart office location and energy usage, preferences, and resulting waste in the canteen. Also, they could be used to predict how they could behave—when doing sales, closing a trade, or working in teams. Even their optimal attraction and retention could be addressed, anticipating, for example, when they are about to decide to leave the company as the competitiveness-to-happiness-to-attractiveness dynamics of other peers (other cyber twin companies) is gaining ground and making scores in the videogame of their industry of reference. Human resource policies would then get optimized and follow a "pay what you need" approach (you just pay what is required to retain your talent and optimize their productivity ratio). The entire human resource-related processes would also get redesigned, incorporating the flow of the information and all the intelligence available on who is, for example, more likely to be successful and make a career, for a given role, in a given company. Finally, cyber twins would also come to dominate the leisure and social world, with real-time analysis of the best match-making available for romantic dates or a new theater production or musical entertainment. Social websites, already powerful per se, would get an entirely new and expanded dimension if IoT become pervasive and with ubiquitous technology to connect them. They would know not only where we are and with whom, but also how we feel and how happy (or unhappy) we are with them, with no need to connect to a website, as feedback would become part of our cybernetic twin and transmittable via other IoT—be them glasses or contact lenses, earplugs, or else.

2.3 Alice's Mirror

In *Alice's Adventures in Wonderland* (Alice's Adventure in Wonderland 1865), this mirror connecting the real and the fantastic worlds works both ways. In the new, meta cities to come, this should also be the case. Everything and everyone in real life can get "mirrored" in the cyber space and in a cybernetic way, but the opposite can also be true, with cyber simulations that are initially tested in the bite world and then mirrored and replicated in the atom one. On one side, the entire "smart" city is feeding information into the real world and the physical twin that helps in optimizing, innovating, and evolving. On the other side, the physical twin city, the one in "flesh and bones," is also feeding information into the cyber one, with plenty of data that are captured offline and analyzed by AI apps to derive further intelligence to help in optimizing, innovating, and evolving the cyber dimension as well. Ultimately, a meta city is able to transcend what it is today by linking the cyber and physical dimensions, in a closely intertwined and encompassing—ideally seamless—way, to then becoming "meta." These meta cities will, in turn, open up multiple, future possibilities—offering a window on a disruptively new cy-phy economy that will require new supporting technology, new buildings, and new infrastructure and, in parallel, new regulatory, political, and social arrangements to make it sustainable. If current cities are likely on an unsustainable path (as we will discuss this later on and with reference to pandemics or climate change risks, as an example), meta cities will need to address all kind of potentially unsustainable issues and offer cunning solutions to some of the structural weaknesses we are experiencing today.

Ultimately these cyber mirror worlds, as explained by David Gelernter, of Yale University, are not mere mathematical representations of the real ones, as they can take a life of their own and act on the physical one in turn. Our point, though, is that the opposite can also be true and would become so in a meta city context, where new competitive advantages could be created on the basis of a balanced competitive exchange between bites and atoms, at individual, company, city, and geopolitical level. This exchange would be based on data, the new "oil," as *The Economist* has suggested (oil is becoming, by the way, a dirty world, in the context of the climate change challenge and the gathering and management of data should also be wary of not going through the same path). Data economy is, however, in our view, an improper name, as data without the ability, through applied analytics of some sort, to mine them and turn information into intelligence is worthless. In fact, more than in a data based economic approach, we believe that value creations could rest on an information-intelligence (info-telligence)-based one. Not even intelligence would then be enough, if we cannot find ways and means, and end uses, to make it relevant and serviceable. In fact, as Robert Nozick has clearly explained, intelligence and even objective truth have little use and economic value if not focused on the delivery of very specific, serviceable, and invariant enough, cunning use cases that are able to crack relevant and critical needs. These use cases need to offer workable solutions, designed and executed to change and improve the way we live and work—to make a positive impact on people and their environment. These solutions should then not only address basic utilities—such as energy consumption or traffic optimization (admittedly, some of them, relevant problems and with a practical everyday impact) but aim higher. They could also address ontological questions and provide sense of purpose to an augmented, space-time framed but also extended (because of technology) human condition. They could support us in reaching our full potential by synergically managing our digital and physical dimensions—offering us an almost infinite mobility, for example, through augmented reality and real-time 3D communication managed via data and AI or via exoskeletons that can allow us to run faster and be stronger—also because of the help of data and AI or by travelling from Los Angeles to San Francisco in a few minutes in a bullet train running in an air-vacuumed tube, also because of data and AI.

We have called this higher ambition as a fully augmented and enmeshed cyber-physical ("cy-phy") world, where we can live in cy-phy, meta cities and that will likely lead to the emergence of cy-phy winning companies (and individuals) that are just much more than "smart." Ideally, as part of this evolutionary journey, cities will become more sustainable, living ecosystems as well, able to optimize the wealth-being of individuals and households. The identification, analysis, and active management of cy-phy twins are a first attempt to stretch and leverage, then reconcile and merge, the cyber and physical dimensions of meta cities with ontological ends. For example, Cityzenith, a US company, offers simulation tools to digitally monitor how the physical city behaves and to test then digitally how different development scenarios would impact the physical world (before being actually built). Multiple applications are then made on offer by its app, from real-time monitoring and remote control of complex systems to scenario testing and strategic planning that becomes dynamic as physical data are fed-back into the digital model and so on, following an iterative process. These simulation tools, typically starting from digital twins and then going in revert, can be applied at full scale at large development or regeneration programs. For example, the Glasgow Future City Initiative and the "from scratch" development of Amaravati (the new capital city of Andhra Pradesh, in India) are heavily reliant on them. Many other ongoing experiences can then be traced down to other international cities, including Singapore, Boston, and Jaipur or (as planned at) the Arexpo in Milan.

2.4 Digital and Physical-Digital

No matter how great basic utilities are made on offer and pitched to perspective users by "smart city" programs, a mostly digital only twin approach would fall short of addressing key ontological questions such as our sense of purpose as professionals and human beings. It would also leave untouched some increasingly relevant, hard, pressing problems such as social inequality and gentrification and the unlevel playing field that is characterizing our professional career and personal mobility in the city—from the exclusive pre-elementary schools of New York and London upwards. These problems, and the ability to address them, would impact heavily on the well-being of people and could lead to estrangement, even revolts, and the ultimate disruption of the fabric of society. A more balanced, personalized approach to managing cyber and physical dimensions to solve ontological questions could instead support sustainable cities doing just that. For example, high-tech innovation has certainly driven further polarization in the distribution of wealth, with huge value accumulations for trillion-dollar companies, still partially owned by a few, incredibly rich, co-founders (and usually still in full control of their governance, thanks to special shares and voting rights attributed to them). In a way, the more the digital innovation happens in highly secured, selectively reclusive spaces, the more it will spur inequality in opportunities, gained abilities, developed capabilities, and shared intelligence and hence in income and wealth. Vice versa, the more is embedded across the city and in public spaces, and starting from public and private platforms mostly relying on the cloud (hence more flexibly usable as on a "pay per use basis," lowering or zeroing the investment hurdles required to tap AI/ML applications), the more opportunities will spread around, supporting cooperation and innovation from the crowd.

For a city to become "smart," it may be enough to offer basic digital utilities—kind of passive, digital services that each citizen receives for free as part of him/her being a taxpayer of that place. But to truly become cyber, and then meta, these utilities need to get personalized and supportive of the citizens work and life projects. More importantly, they need to allow interactions and feedbacks and, ultimately, offer the opportunity to access, at reasonable economic terms, the central pool of digital and physical data captured (the "oil" that should, at least partially, be co-owned by the meta citizens) and the array of AI/ML applications that are safely stored and usable over the cloud. The renewed focus and relevance of the physical dimension could also help in rebalancing the wealth accumulation equation, now strongly in favor of a bunch of digital platform "barons." Even more so, if citizens are made aware of the potential value their streams of physical data have and if, on top of some reliable pricing tool, they are also trained, instructed, and supported on the way they may be monetizing them as best, in exchange of money or else and in accordance with their overall, larger pursuits in life, we may want to sell data for money, or to access computing power and AI/ML tools, or to receive free education and training or health care monitoring and first aid support. We will discuss later how this marketplace for data could develop in efficient and effective ways and by creating new sorts of financial (information) intermediaries—but the important point in here is that, whatever the mechanism of optimal exchange and allocation of big data and AI/ML computing capabilities, the goal of a meta city should be to make it "open," transparent, easily accessible at reasonable costs, meritocratic, but also compassionate and promoting its utmost inclusiveness.

Inclusiveness can, for example, start from very obvious, hard problems—e.g., from basic issues such as ensuring digital access and prowess to tech illiterates, starting from elderly people and from the poorest cohorts of the population, which may have not had the opportunity to own a PC or smartphone, or to receive any formal training on basic spreadsheets, not to mention basic coding and other tech-savvy applications. A sustainable city should then be promoting free access to open data sources and the simplest usability of whatever it takes to make citizens as digitally included as possible. Inclusiveness may however relate to the physical dimension as well, such as when ensuring improved and then full accessibility for people with disabilities (Better data paves the way for improved accessibility 2020). Cyclomedia, for example, is improving accessibility by applying its imaging of physical spaces to create a detailed view of the city's roads and pavements. It then combines 3D mapping technology and advances in artificial intelligence to provide data for the city to make urgent repairs to pavements and to create an investment strategy for developing accessible streets in the long run. Following a "profiling of one" approach, dynamic customization and optimal designing could also foster inclusiveness and accessibility and could be applied to buildings and homes, the obvious other physical element of a meta city. High-harc, a US-based start-up, is also using iterative design processes to create tailored 3D building plans and where algorithms do the final modeling (Automation finds home in building design 2020) to optimize endgame usability.

2.5 Socially Acceptable Hence Sustainable

However, digital and physical inclusiveness is not enough, if not supported by a common and reliable platform that can ensure trust, fairness, and transparency to all the people included, on the basis of shared social value and of a democratic infrastructure. In fact, to make a meta city truly sustainable, this needs to be made socially acceptable. This means, in turn, that the city will need to ensure that the wealth it will create will not be even more unequally distributed than what is already happening today, avoiding the risk of a hyper polarization where the owner of the best algorithms can end up "owning it all." In order to do this, careful considerations will need to be put on the governance of data and intelligence. If data is the new old (or gold, or rare earths, or COVID-19 vaccine, or whatever the scarce and most valuable resource can be), it needs to be priced and transacted effectively and efficiently and avoiding market corners and oligopolistic or monopolistic market structure. Some sharing of this data needs to be made available and "refined" (via AI/ML applications) not only in centralized, dedicated spots but also in an open source format, as an asset and as a process that are both to be considered as "public goods"—critical to unlock innovation opportunities and their creative destruction waves that will in turn impact on the overall economy.

Following a first trend, data centers have been centralized and run over a single cloud hub, packed with servers storing data and crunching information. An emerging trend is then now suggesting a more diversified approach, with multiple data centers, both public and private and multi-cloud hubs spread across different locations, to make them not just more flexible but also more resilient and with extra buffers—should anything major happen. Diversification could, for example, consider international hot spots as buffers (in case of catastrophic events hitting the city of reference) but also multiple cloud hubs around the city area, to take full advantage of the adoption of 5G technology—where data are getting processed more and more "at the hedge," closer to where they are collected and then meant for end use. This makes the system more flexible, personalized, and agile, with "capture to use" elapsed time cut to a fraction. A number of different approaches are still in motion and competing with each other and with mixed results. Some meta trends could be identified, but likely, mixed approaches could also prevail in the future. For example, data have been initially gathered and managed in pretty rigid and structured data bases that—be them public or private—were setting the rules for archiving and retrieving and following traditional categories (more focused on the sources or taxonomy of data then on their end use). Then, they have been captured and hosted into more unstructured "data lakes" (with structured data base "rivers" flooding their data almost with no rules) that have become the closely guarded unrefined oil reserves of firms.

Data lakes have been initially built internally, at the level of individual companies, and have then been reversed on the cloud, for the lack of skills of small enterprises and to leverage the obvious advantages coming from scalable investments in technology. Larger and larger data lakes have also been pursued (ideally at cross-companies level, a task in practice almost impossible to pursue given regulations on data ownership and sensitivity on confidentiality and data as a competitive industrial asset) in order to also leverage greater and greater amounts of data—differently from tanks of oil, the greater the data available, the better the chance to use them to derive superior competitive intelligence. Given the sensitivity of such intangible assets, data lakes have sometimes been forced to re-localize, as supranational regulators and governments has well have been trying to protect the data owner, the personal dimension of them, and the competitive and national security aspects of this. Governments are then now trying to assert digital sovereignty as well, demanding that data are not leaving the country of origin and be subject to safety standards and personal identity protection laws—not to mention national defense policies, in case of data with a potentially military application. As the threat of a new American-Chinese cold war is emerging, this sovereignty aspect could become more and more critical, with data that are subject to greater international constraints, scrutiny, and interference of policy makers. All this would constraint the ideation and evolutionary power that is implicit in data and semantically articulated information and that is just magnified by an "open data" approach. This, as we will discuss later on, would instead greatly benefit the citizens of "cy-phy" sustainable cities that are pursuing sustainability, inclusiveness, democracy, and freedom in a peaceful environment, where human intelligence is harnessed and shared for the common good.

2.6 Harnessing Data for the Common Good

Harnessing data as a first step to then harnessing all human intelligence to overcome the most critical challenges ahead of us is certainly a visionary task, and not short on ambitions and technological difficulties—as articulated by the inspirational CEO and co-founder of SparkBeyond Sergey (Sagie) Davidovich, an AI-based platform providing solutions that are also leveraging all open data sources available on the web. That's why discussing data bases, cloud solutions, and their governance rules and potential constraints is so important—differently from oil, as humankind, the more data we share, the greater the ideation and intelligence potentials we should be able to achieve and—hopefully—the higher the chance not just of addressing common threats such as pandemics or climate change but also the risk of internecine wars. We will argue later that an "open data" approach is therefore beneficial for the overall mankind and, no matter what the tortuous and bumpy journey, the way the all info-telligence industry will need to go—with massive value migrations then flowing from data owners (also said sometimes data "barons" or "robbers," if they have taken data on an unfair trick) to intelligent solution providers—players that are able, as in the aim of SparkBeyond, to support ideation leveraging technology on top of human ingenuity, and the creation of serviceable truths—also derived from mostly public data—that can have a beneficial impact on societies and markets, and supporting the evolution of civilization in sustainable cities.

If this is the end vision (or dream), the starting point needs to be in any case the right setup of ownership rights and governance rules, for the optimal and fair appreciation of what data could mean, in terms of potential power, monetary value, and else. With no fair rights and transparent governance, there cannot be an efficient and effective and, most importantly, fair and inclusive market for data—providing the right set of incentives to participants to do better and do more, to further augment the harnessing of all humankind intelligence. And with no efficient, effective, and fair market, there cannot be a sustainable evolution towards more developed forms of pooling and sharing of data—that is the best starting point for the all-out intelligence harnessing effort envisioned, with the help of AI, by Sagie Davidovich. Once ownership rights and governance rules for data and their end uses are cleared, they can become transactionable at a price—also based on their demand and supply, at the whims of the end customer and of the original owner. They may then be regrouped into data warehouses (adopting more structured approaches in archiving and management) and lakes (going instead for more flexible and unstructured approaches) and be mined there by AI/ML apps that aim to find serviceable and invariant enough truths that can be sold at the highest price for their most productive uses—not dissimilarly from the refinery process for oil.

This mining (refining) process can be done at the level of a single, centralized cloud or across a multi-clouds network, by the owner of the data or by third parties (the two may, of course, coincide). In an "open data" visionary scenario, all kind of parties could have access to nonsensitive/nonconfidential data that would be treated as "common good." Winners would then prevail on the basis of their ability to support ideation from data and create superior intelligence starting from the "common good." Everybody will be given a chisel and block of marble, but only the Michelangelos would be able to distinguish themselves, reinterpreting the row material to incarnate an idea of truth that is so cunning and objectively recognizable to stand as "beauty per se." Michelangelos could, of course, be talented individuals but, more likely, team of talented individuals working for AI companies. Data mining platforms such as Snowflake, for example, can already work on data and stretch their applied analytics across different computing clouds. Other specialized cloud and data base players can also develop their offering and focus on capturing data in real time and via digital streams (e.g., leveraging 5G technologies) and offering AI/ML as part of the service along with other, adjunct services—as it could be the case for Google Cloud that has teamed up with TIM (a telco provider in Italy, also owning Olivetti, an IoT provider). Smaller players such as Databricks are also offering an AI platform available as a service, completed with tools to cleanse data, build and test algorithms, and deploy them. And, as mentioned, SparkBeyond is offering to use AI starting from one of the vastest data base existing as of today, having linked all the open data sources available over the worldwide web to its AI engine, mostly based on genetic algorithms that are used to generate millions of ideas per minute, rapidly testing, selecting, regrouping, and ranking them on the basis of their relevance (kind of cause-effect correlation power to explain the dependent variable being investigated and hence address the main question requiring the harnessing of global intelligence) to then make a valuable impact on markets and societies.

It should be now clear enough how all this process of data capture, management, optimization, and use would be constrained by any form of digital divide—at city and company level, not to mention at country or geopolitical blocks level (think of a Western versus Asian divide, with Africa lost in between). Most of the opportunities to optimize resources and make cities more sustainable and better livable would fall short if data silos are created, without allowing their cross-reference to look for root causes and potential impacts. Not only greater and better intelligence can be created if more data are used but taking out entire sets of data would sub-optimize, if not impair, the entire info-telligence production process. This could well happen as well at the level of the organization, where different divisions are afraid to relinquish power along with their data and relative fiefdom of information—and this could limit and impair the overall company and economy's competitiveness. In short, any limitation to the full use and transactability of data (at market price, or following some other rule-based, democratic, and libertarian approach, with certain data sets that are considered as "public goods") would limit the inclusiveness and efficiency/

efficacy of any kind of ecosystem system and make them less sustainable, less socially acceptable, and less competitive in the mid-long run. An "open data" approach would instead make them more easily available and fungible, tradable at market prices or at a given, set price when classified as public goods (some of them could be made available for free, in normal times or during extraordinary emergencies—as for the COVID-19 example). An "open data" approach would then require, but also support, the development of the citizens' intelligence on the optimal value and best uses of data and of their resulting intelligence. Collectively, they could take a greater ownership to develop their own sustainable city, contributing energy and passion to co-development, assuming responsibility and contributing creativity and new ideas on the best use cases as discussed in a democratic, open-crowd-based debate. Specifically, the mix of cyber and physical data could be collected and analyzed to form a view on local activities and the behavior of the single individual and larger teams and of small and large companies, to offer multiple opportunities for social inclusion, wealth-being creation, and preservation of the environment and for the pursuit of a better work-life well-being. Ethics would then play a critical role to ensure the build-up of trust and ensure democracy and freedom in normal times, but also the execution of quick and effective measures (temporarily limiting the freedom of the individual) in times of need.

2.7 Smart, Cyber, and Meta Cities at the Time of COVID-19

A meta city, mixing up the best of the cyber and physical dimensions and able to derive more cunning and tailored "cy-phy" solutions to address urgent needs and issue of the utmost relevance, should also be able to qualify as "sustainable," in normal times (e.g., with reference to its contribution to the stabilization of the environment and fight of the climate change related risks—discussed later) and even more so in extraordinary circumstances, as at the time of the COVID-19 global outbreak. Think of our pandemic recent, dramatic experience. This has led big cities and even entire countries into a prolonged lockdown, with most people working from home in improvised "smart working" setting and with entire families relying on Amazon-like ecommerce websites (and offline logistics-based, physical distribution) for almost everything that is not food or medicines. It has led us to rapidly changing our ways of working and living and to adapt to the new lockdown context, relying more on digital communication and less on physical mobility, but also on alternative (and physical) supply chains that can support the cyber dimension: allowing us to be virtually present and with a physical backup when required. Progressively fusing—while holistically reinterpreting—the dimensions of communication and mobility and of transportation and experience, health preservation, and well-being. A meta, sustainable city could in fact be something more.

A "smart city," digitally hyperconnected and supporting basic utilities, could have allowed us to work and live better through the crisis, with greater cyber security and higher performance connectivity. For example, Singapore, likely the most connected and advanced city globally, has been able to better contrast the diffusion of the virus in its early stage and thanks to technological applications used for mass surveillance and digital health checks and first aids. But it has then fall short, allowing the resurgence of multiple waves of the virus, when it has become apparent how the very high safety standards and services applied to its wealthiest citizens were not matched by something similar for the numerous immigrants and poorest, temporary workers that are still employed by maybe the wealthiest country (ad measured by GDP per head) and certainly the most hyperconnected in the world. The success shown in contrasting the diffusion of the virus has also be evident, at scale, in China where, on top of tracking smartphones, AI-powered face recognition technology (trained to identify people also when wearing masks) has been used to ensure (and sometimes brutally enforce) the lockdown of entire cities and to reconstruct the network of recent contacts and relationships of contagious patients. Also, in this case, success has fallen short in the early stages of the outbreak, when a more democratic and more open media and communication system would have allowed the transparent diffusion of relevant information without cover-ups.

Instead of a "smart city," a more evolved "cyber city," extensively connected by the way of IoT, could have worked even better by allowing, for example, personalized telemedicine and continuous monitoring of basic health functions (including body temperatures, cough, and heartbeats—that could be monitored within a flat or building as part of their enhanced, personalized domotics). A cyber city could have allowed the screening of the ways the virus is extending and developing as epidemic and then pandemic, if our basic health functions are then associated to a mac-address or an IoT-connected bracelet and then followed as we walk, drive, and fly. And a meta city, or better, sustainable city, that is also supporting the inclusiveness and extended coverage, with digital and physical services, of all kind of people—resident citizens and immigrant workers included—could have prevented some backlashes; and an open and transparent sharing of information, carried over by a democratic and participative system (also key part of our proposition of sustainable city) would have also supported a more immediate and concerted effort, within single cities and among countries and geopolitical blocks. A number of more granular examples could be easily made.

On detection and traceability, for example, the network of people we are getting in touch with could be registered and analyzed for interconnections and cause-effect relationships—to better understand and contrast the diffusion of this and of new viruses to come. Israeli police have apparently done this effectively, tracking and checking quarantined people, based on the geolocation of their smartphones or smartwatches. And Chinese policy makers have been using face recognition in public spaces (including when wearing protective medical masks) to monitor the flow of people and track potential carriers of the virus (leveraging a technology that is able to measure body temperature at a distance and when we are on the move). All these check points and extended diffusion of data would work on both ways. On one side, data on people's movements, basic health functions, and behaviors would feed the cyber twin city, built as a mirror of the physical one, and across basic, common utilities and more sophisticated, personalized services—with feedback mechanisms to ensure the cybernetic dimension of it. On the other side, cyber data (and mined intelligence) would flow back into the atoms-based physical reality, with just in time advice to people on how to behave in a safer way, signaling when symptoms are detected and suggesting the best way to keep working and living in the most productive and enjoyable way—given constraints.

The cyber city would then get optimized, by running scenarios and understanding how the pandemic crisis could be better dealt with, and the info-telligence derived could then be reversed into the mirror physical city, with the ideal solution provided ready at hand—to avoid the trial and error approach applied by policy makers across the world. Not to mention that, given the inherent un-linearity of the typical development of any epidemic or pandemic crisis, the use of structured and unstructured data, in real time, is worth much more than the value of long historical series of structured but also old data that are referring to different ecosystems and biological environments. And that AI/ML apps could be better in spotting early patterns and in sorting out emerging trends than traditional statistical, conjoint analysis—using heroic hypothesis on the correlations' stability. Ultimately, the search for the holy grail of a universal vaccine able to address once and for all the virus would benefit largely from this big data and AI/ML-based approach as well. Leveraging the parallel, mirroring dimensions of cyber and physical, larger and quicker numbers of vaccine testing could be done, reducing the approval time and go to market elapsed time and likely suggesting the optimal way of prioritizing the rollout of any available cure, certainly based on people's age, but also on their mobility, lifestyle, and risk exposure, not to mention specific genome, when this will become available at individual level, hence supporting the "medicine of one" managed by sustainable cities.

2.8 Fair Use, Fair Trade

The development of a global pandemic can certainly and temporarily allow for some greater latitude, in terms of use of data and personal information that can be captured, analyzed, and used ex lege. Multiple use cases, performed by the relevant authorities, could ensure, for example, that "social distancing" is performed by customers when visiting shops (with fines for customers and enforced closures of the shops to follow, if this is not ensured) or that people avoid get-togethers when staying in public spaces or do not travel outside the allowed zone or perform strictly required jobs, if mobility and work restrictions are enforced (also punished with fines and even forced quarantine). Actually, the recent development of the coronavirus pandemic has supported an unprecedented acceleration of technological applications that are able to do just that, with a greater and greater focus on the mingling of cyber and physical information that are able to feed and intertwine digital and "flash and bone" twin cities—feeding back continuously each other on modeling, behavior, performance, and management. This is however, and rightly, posing a number of worries as the potential threat of a cyber-physical big brother could now count on the maturity of underlying technologies and of (as of now) almost fully tested processes—as they have been applied across entire cities and countries, across the globe—at massive scale and for the first time in history. Not only that. Leaving aside the emergency, pandemic "stress testing" of this enforced monitoring and controlling exercise, now that our online and offline data have been comprehensively mapped, traced, and positioned for continuous collection and analysis, it could happen that—even in normal times—this cyber-physical extension becomes then the new battlefield (or oilfield) of few, dominant "big digitals." They could be coming for our faces and being able to read into our emotions and inner needs or absconded thoughts—a notoriously threatening use case would be the ability for North Korean leaders to be able to read the minds of their people, when Kim Jong Un speaks to them.

Given the explosion in digital tracing and facial recognition capabilities and with sprawling, more and more extended infrastructures, able to potentially deep monitor us and 24/7 (in private and public spaces, all connected in real time and able to recognize us and read our mind and hart through our behaviors), international legislators are putting a lot of effort to design a legal framework. This should prescribe and rule if, where, when, and how to use the data and intelligence produced from this cyber-physical explosion of data capture and mining capabilities (Artists and activists offer privacy hope as facial recognition spreads 2020). Our identities are already being captured, digitized, indexed, clustered, and analyzed for correlations and relevance—resulting, more or less sophisticated intelligence is then packaged and sold, with and without our formal (when, e.g., asking us to sign in on a 45 pages of non-negotiable contractual terms or suffer digital isolation and marginalization) or substantial permission (when actually understanding the real implications of it—we trade our information for something substantial, such as a free service). To avoid the worst of it, and pursuit the best, innovative technologies to support the development of sustainable cities, the fair use of data and of their derived info-telligence need to be ensured, regulated, and grounded on the principles of inclusiveness, democracy, and freedom. It is of course potentially arbitrary to define what a fair use is and could be, under different scenarios (e.g., the "fair use" of the information regarding our mobility is changing during an enforced city lockdown), even if a set of basic principles and rules can put limits and unsurmountable legal constraints. But at least a proper debate and some kind of public and private framework need to be defined.

A fair use, in economic terms, could, for example, be derived, within the limits set forth by the law, on the basis of a fairer trading system, between the original owner of the data, the wholesale buyer, the storage, computing power and applied analytics supplier, the reseller, and the final end user. The definition of what a fair trade and a fair transaction (or multiple transactions) and hence fair valuation of the monetary price of data and derived info-telligence is not, by no mean, a small challenge, as the price of data would need to change across the info-telligence extraction, manipulation, and production to distribution value chain and given their final end use (which may be multiple and completely exhaustive/completely exclusive or not) and certain characteristics that makes data unique. Data are in fact a very peculiar valuable and scarce resource (even the term "scarce" could be debatable, given the exponentially increasing throve of data now made available). For example, data are not depleted with use and could be reproduced an infinite number of times, at almost zero marginal costs. Also, data are valuable, but the potential info-telligence that can be extracted by them could be many times as such and its function could change with volumes—in likely unlinear and reversible trends—and because of their different combination. Certain data are very valuable as long as unknown to most of the relevant population as well, hence need to be protected as confidential; and, the value of data can change rapidly in time as they may be subject to rapid obsolescence or, vice versa, be valuable only if they are available for long time series, etc. In short, a fair use based on free trade is not just an incredible challenge from the perspective of the law and regulation, but also from a scientific, methodological, and practical point of view.

2.9 Big, Digital, and Smart

Big digital players, usually defined by a business model based on a big data platform aggregator (e.g., a là Amazon, Google, Facebook in the western world), have been capturing most of the value coming from the emerging data and info-telligence economy and across multiple industries, with significant economies of scale and scope and polarizing network effects. Some of them are now worth trillions of dollars, and their rapid growth has been paralleled by the (negative) value migration observed for other traditional players acting in other industries, from retailing and consumer goods to finance. Already dominant in the pre COVID-19 world, they have become so even more after the global health crisis. At mid-June 2020, just when western countries were re-surfacing from the first wave of the coronavirus, the top 100 companies "prospering in the pandemics" were largely populated by them, with Amazon on top (+$ 401 Bn in market capitalization), followed by Microsoft (+ $ 270 Bn) and Apple (+$ 219 Bn), then Tesla (in a way, also linked to data and AI, with +108), Tencent (+93), Facebook (+86), Nvidia (working in digital videogames, +83), Alphabet (+68 notwithstanding the global collapse of advertising), PayPal (offering online payments, +65), T-mobile (+60), Pinduoduo (+55), Netflix (+55), Meituan Dianping (ecommerce, +54), Shopify (+51), and Zoom Video (+48) (Prospering in the pandemics: the top 100 companies 2020). Zoom may be in fact the most inspiring example regarding our changing ways of working and living, with its work-from-home and socialize-from-home proposition and making its fake digital backdrops a folkloristic and cultural touchstone of the crisis lockdown.

Provocatively, we could argue that most digital players started their business and incredible value creation journey by capturing data "on a trick" and by betting on the most likely future scenario to come (a scenario of extended connectivity and digitization of life at first and then of progressive fusion and enmeshing of the cyber and physical dimensions to come—something that was already on the cards and came just as brutally accelerated by the pandemics). On one side, these big digital players have certainly brought to the masses incredible technological innovation and offered unprecedented value added. On the other side, they have been offering these value added for free or for a small, subsidized price—these investments have been recovered by capturing data and information, not just for monetization through advertising but also sometimes in unobvious ways and for often undetected and little-known (and sometimes malicious) end uses (e.g., the Cambridge Analytica—Facebook scandal, when personal data where used, without consent, for political advertising). Leaving aside the improper use of data, a number of great success stories can be traced from that basic "trick." Google, for example, started its own business by offering a great search engine, but has since been using this highly analytical tool to rank offers and prioritize and promote, with tailored advertising (for a fee) certain products and suppliers instead of others. Also, it then started offering free email and storage, but it can read into our email and offer personalized advertising in almost real time (also for a fee). Contractually, what it does is specifically agreed by the end user and in accordance with the GDPR (General Data Protection Regulation) in Europe. This is happening, for Google and many other web service providers, when we are asked to accept at long and tricky "terms and conditions" agreement of use, which are un-negotiable and likely impossible to understand in full for a normal user without the help of a set of specialist lawyers.

We could, of course, decide not to use Google, and, assuming we can manage with the quasi automatic redirection of our smartphone browser towards it, we would end up on a similar big digital player, like Yahoo!, for example, with very similar "terms and conditions" attached. Eventually, we could just decide to opt out from any search engine, or social website, or emailing and data service hub and become a derelict member of the society, left alone and unconnected—a definitive looser in the battle brought forward by the digital divide. Not included and not socially accepted—hence unable to compete at work and in life and hence to sustain our wealth-being. This is not to say big digital players have not contribute greatly to our wealth and well-being—they have certainly done so, and, going back to the case to Amazon that has benefitted most from the pandemics, it has done so not just because it has kept delivery on its "great convenience/great value for money" proposition, but also because it has kept offering and delivering indispensable products, serving billions of people in a moment of great stress and almost providing a quasi-public and quasi-social service. But, leaving aside for a moment the technological innovation and value added they bring—something we would never be ready to give away now that we are used to it, in normal and even more so in pandemics time—they all pursue a very simple business model, acting as an intelligent and more sophisticated broker dealer between end customers and vendors.

2.10 Data Traders and Arbitrageurs

In fact, they trade and arbitrage on data and personalized information. Eventually, they are also able to capture our attention and to keep us staring at a small screen for hours at end. And ultimately, they are great in building extra value by mining data via applied analytics and AI/ML to produce intelligence as well. They may then decide to retain this intelligence for their own use or for the sale to other end users, typically corporate, but also policy makers—to help them sell us other stuff, designing "marketing campaigns of one" and reaching out individuals with the right product at the right time, often knowing in advance, even before we, as consumers, know it, what our next need or wish could be. If data is the new oil that they extract, transform, refine, use, or distribute, they are then able to capture value from this on multiple basis, leveraging their being "big," "digital," but also "smart." As a start, it can be assumed that they are able to trade rough data at below its fair value—e.g., they typically do not pay money and do not share revenues with data owners but just offer them other frills or free services and not on a fair trade basis—hence the first wave of value accumulation for value capturers and resellers. Overall, similar players (because of their size and of the scale and network effects driven by their platform-based business models) could end up acting in a quasi-oligopolistic way—across global markets where no independent and reliable appraisal on the value of data is available. Even if some kind of fair value on data could be available, it would be difficult to enforce it anyway—when we are asked to sign in and cede data for a song, given the strong attractiveness of the main digital players and the limited offer of comparable, alternative suppliers. In fact, a fair price can tend to be reflected in an observed market price when there is some negotiating balance between demand and supply (and competing suppliers) and no market cornering is possible or allowed in highly liquid, transparent, and fairly regulated setups.

On top of this first "value migration," big digital players can then add a second wave of value creation to their own benefit, as they develop analytical capabilities to extract superior information and produce cunning intelligence. Even assuming they are following GDPR, they create intelligence on specific cohorts by segmenting the population into homogeneous clusters that are good for serviceable, well-defined uses. They can also use anonymized data at individual level that can then be used for multiple purposes, including tailored R&D and with potential, profiled outcomes and proposition that are sent back to the original individuals, upon their contractual sign-in and legal permission to receive feedbacks and potential solutions. Solutions on offer could regard anything that is valuable to us and extend to all kind of economic and social dimensions: from our financial risk/return profile, to our way of living and consuming, to our desired lifestyle and political whims, and to our physical/psychological health and ontological dimension. These solutions would not necessarily entail the delivery of a service or product "owned" by the big digital players. In fact, most likely, they would instead pursue an optimal match-making between demand and supply, by either selling these information directly to vendors or, more often, working as their broker in a "many to many" competitive digital information marketplace, where they gain a fixed commission or a share of the revenues of the ultimately transacted goods or services. Eventually, they could also use data and intelligence to develop their own solutions, to address pent up demand and internalize some R&D, design, and production-to-distribution capabilities (e.g., Amazon offering its own groceries or self-publishing platform; Google and Facebook entering payments and deposit gathering).

If these may look scary, even the upcoming, widely anticipated "summer of Techlash" (Moves to limit Big Tech still only half formed 2020) may pose some worries. A flurry of investigations and proposals, aimed at limiting the power of Big Tech, has already been running for some time, with geopolitical clashes between Europe and the United States, and between the western world and China. Anti-trust investigations are already running on both fronts of the Atlantic, but with no similar action happening as of now in China; this is just promising to create a further unlevel playing field in the Klondike-like "gold rush" on data and AI. In fact, the single biggest motivation that is bringing to some action in the western world—the desire to tackle the power of big digitals in acting as gatekeeper to data, information, and intelligence may be exactly the end-game in sight of China, which is open to allow private profits for its own big digitals (e.g., Alibaba, Tencent, Baidu, JD.com) if this is the price to pay to keep controlling and monopolizing social (and political) discourse on the web and extended media. The battle to prevent the spread of fake news or abhorrent material is also part of this game, but again leaving ample space for the government of the country to decide what fake news or abhorrent material is (the recent case of Hong Kong is an interesting one) and is actually posing another and maybe much larger, questions. If big digitals are acting as dominant platform and "gate keepers," what about if the rules and regulations that are aiming at preventing the unfair exploitation of their dominant position are instead used by the very same rulers to create a faked sense of democracy and justice?

2.11 Fair Digital in Fair Cities

Likely, the best answer to the main question posed by this chapter, on the ways to make the new, upcoming "cy-phy" future fairer, along with a fairer economy and society, may lie within the design and management of the "cy-phy" cities to come. The ownership, capture, management, access, and elaboration of data will increasingly be (we have discussed) a source of wealth, well-being, and power. All this could allow massive value creations, for the benefits of the masses, or the incredible accumulation of riches for a few—likely exploiting value migration and quasi-monopolistic, gate keeping business models. And could also allow the formation and dominance of populistic political forces in the western world (and of dictatorial ones in other parts of the world), able to manipulate the whims of the people, telling them what they want to hear, manipulating their decision making by providing fake news, leading them into a matrix-like limbo where they live in a "twin reality" that does not actually fully exist or does appear in a very different format vis-à-vis what is in its full actuality. Likely, we said, the best answer to the threat of private dominance by an handful of big digital and to the one of public ones by a few policy makers and dictators lies in the "cy-phy" cities if they will be able to develop as inclusive—closing, for example, the digital divide or the gender and race inequality and driven by an holistic pursuit of wealth-being for the citizen in full respect of the environment and of the basic rules of democracy and freedom. This is not a small challenge though and will require all the ingenuity that big, successful cities have shown to possess in the past, polarizing the creation of wealth, the attractiveness of best talents, and allowing for continuous self-rejuvenation and regeneration. But before going deeper into that challenge, we should maybe consider a more basic one, regarding the future of cities. Do, in fact, big cities have a future at all?

The question looks counterintuitive after so many statements regarding the trend of urbanization (with 75% of the population predicted to live in cities by 2050), their polarization in terms of contribution to GDP creation, and their attractiveness leading the way in the competitive battlefield where countries and entire geopolitical blocks are combating every day. It certainly looked stupid at the time of the beginning of this work, in a pre-COVID-19 era. But looks less so, in a post-COVID 19 one that is seeing the revenge of rural areas, with people fighting better pandemics in the countryside, working remotely from more pleasurable seaside places and from the mountains. With great connectivity becoming luckily more extended and with the promise of 5G almost real-time synchronization, with 3D printing and augmented reality and the convenience offered by cy-phy players like Amazon (that can deliver to your home basically everything, within 24 hours and for a negligible fee), what would be the point to live in the city anyway? Why paying € 10 or 20 K per square meter for a small, old, cranky flat in New York when you can work from your second home in Santo Domingo or in the lake region, for a fraction of the costs and much greater quality of life? Why spending so much time commuting every day to the office (even a ride from Notting Hill to Canary Wharf could take one hour), to work in a small office or a noisy open space for 10 hours, with the risk of infections, accidents, terrorist attacks, and what else. What is the point then to be in the city?

This is a critical question to answers, with impacts, coming from our future preferences and behaviors, likely to change the global current wealth, well-being, and power maps much more than the COVID-19 itself. It is a difficult question as well, as we walk along downtown streets that still, even after lockdown, almost deserted. In fact, the coronavirus has struck mostly big global cities, from Milan to London, from New York to Sao Paulo—most dynamic cities seem to have suffered disproportionally more (After disaster 2020) with rural areas taking some revenge because of their better safety, well-being, and even higher productivity when you are put at work on something brainy (if doing smart-working—this is, at least, the experience of the author of this book). Cities have been successful because they are able to bring together most talented, vibrant people but also because they offer theaters, shopping venues, restaurants, museums, and night life and lots of well-paid jobs opportunities. US cities with more than 1 Mn people are 50% more productive than the others, according to *The Economist*. But will this remain true in the future? James Gorman, CEO of Morgan Stanley, has stated that COVID-19 has shown that his bank can operate with (almost) no physical footprint. Jess Staley, CEO of Barclays, has also been wandering if 6.000 people-crammed skyscrapers in financial centers are already a thing of the past and just too unsafe to be true (The destiny of density 2020). Not only have big, global cities been suffering more from the pandemics but, also as a consequence of this, they will suffer more drastic budget cuts and with less money to invest on digital technology and regeneration—not to mention the huge physical infrastructures they need to keep hosting and moving efficiently millions of people at all times of the day. You may stop investing in roads, public transports, and elevators in the countryside for a while, and nobody will notice. But what about a city stuffed with people that are always on the move and with a concentration per square km that can be ten- or a hundred-fold that of the countryside? Cities have been left largely untouched by terrorist attacks, pollution, and financial and real estate crashes, but COVID-19 could have a much larger impacts—not only because it is imposing the largest global economic recession in modern history but also because it is potentially changing our habits of living (or not living) the city and its buildings, think of empty offices and lonely shopping centers! Is then COVID-19 the beginning of the end of cities?

2.12 Death of Cities?

As famously stated by Mark Twain, "reports of (cities') death (may) be greatly exaggerated." Certainly, our home is becoming more and more the new turf for residential living innovation, with a room dedicated to office, or shared-serviced, professional rooms that can be hired by the hour as for WeWork likely becoming a feature of new residential buildings (on top of the gym and SPA). Certainly, this can also allow for a more widespread approach to new developments across the land—assuming fiber and 5G are ensuring the high connectivity of the place. And, certainly, the becoming, apparently unstoppable geopolitical competition among main, international, cities may also change direction and reverse, in the short run, with more domestic competition between, say, Milan and the lakes or mountain regions (where second homes are, and becoming more and more lived in) or Rome and the seaside (also becoming very popular during the lockdown and anticipating the summer season "en plein"). Almost paradoxically, a few months pandemic has been impacting on the acceleration of digitalization more than many years of teaching and preaching—high demand is finally pushing a speedier deployment of connectivity infrastructure but also the use of applications, like Zoom, Webex, or Teams, which were already alive and kicking well before the coronavirus. Is then digital causing the accelerated demise of the "smart city"? Eventually, if citizens are smart and digitally well connected (and behavioral norms and companies' rules allow it), they will all decide to avoid the daily nerve-wrecking commuting. Most "smart" citizens will elect to do more work from home and from the countryside, where real estate is cheaper and with larger public spaces, air quality is better, the environment looks great, and you can be 5 minutes away from (real) surfing or mountain biking, since you click "disconnect" from your last videocall of the day.

We feel that it's too early to write off cities entirely and bet against London and New York, Milan, or Berlin—cities have proved to be incredibly adaptive to changes of historical proportions—they have managed world wars, famines, and pandemics that have been much worse than COVID-19. But what is changing vis-à-vis the past is however the availability of digital capabilities and the now widely and competitively available "cyber dimension." Hence, while "smart technologies" could actually accelerate the demise of cities, a new approach based on the recovery of the physical dimension and its enmeshing with the cyber one is what is required for cities to develop and survive, change, and thrive. The buzz of cities is unique—it could be even better if it mixes the cyber and physical dimensions. Physical and intellectual excitement, the search for leisure and sociality, passionate energy and keen emotions, and the willingness to live our lives in full cannot realize at best when we are all living from different places—dialing-in for party from Courmayeur and San Tropez. For business, a more varied cities-to-countryside approach will likely be part of the solution, with less traveling and more staying—with the collapse of distance followed by the augmentation of the on-site experience (think of all the bankers flying from main European cities every Monday to London and returning on the day—they may decide to fly once a month, but for a full-week stopover). For leisure, we could also decide to travel almost every weekend to the seaside or mountains for short bites and do it once per month, but for an entire week. Offices and commercial and residential buildings should then get ready to host, almost by design and certainly with a bit of planning required ahead, a 50% or even 25% world—as we will spend just half of our time in the office when in the city (assuming we already spent, pre-COVID and holidays included, a good 25% of our time out of cities).

This is not going to be all bad for cities as well. As noted by Simon Kuper (2020) the COVID-19 has pushed the adoption of the likes of Zoom and may have made congested cities like Los Angeles more attractive by letting residents cut down on driving. Looking back at history, place like New York has thrived after, for example, the 1918–1919 flu. Richard Florida, an urban theorist, is then predicting a new season of growth for cities—but with younger populations than before, more and better, larger housing and fewer offices and shops—and ready to take a flight when they want to take a full week off in some remote location. This in turn, should lead to a further polarization in the fortunes of cities, with few global ones able to adapt to the new environment and act as "cy-phy" catalysts and many other in the middle likely to face irrelevance—if it's not a super-buzz business-leisure city place nor a super-nice countryside for remote working and living, why should anyone be keen to buy or rent there? Think of the hinterland between the city center of metropolitan Milan and the alps, lakes, or seaside coastal "boutique" cities—what would be the point of owning or renting anything in between? Also as a consequence of that, according to Florida (Simon Kuper 2020) young (and hence more dynamic) people will keep flocking to cities to have fun and make the personal connections and networks that drive careers—with big cities office and retail space falling empty and be converted into homes—likely with greater shared services and spaces as well, and ideally with greater mix-ups of races, genders, sex orientations, nationalities, and religious beliefs (this should be easier for younger people). All in all, it is too early and likely wrong to write off cities entirely, but huge value migrations in the real estate space and across cities are not, in our view, "greatly exaggerated." As not "greatly exaggerated" is the critical and urgent need for cities to rethink themselves, to become more "cy-phy"—inclusive and sustainable—and leverage this apparently impossible symbiosis between digital and real life to rethink and redevelop their special and unique proposition to answer to a very simple question that, right now, should be in everybody's mind: "after COVID-19 and with 5G all over us, why on hearth should I be living or spending time, and buying or renting, just in here… of all places?"

References

Is Silicon Valley's quest for immortality a fate worse than death? Adam Gabbatt, The Guardian, February 23rd, 2020.
Can you become "virtually immortal"? A Silicon Valley start-up thinks so, Dave Schilling, The Guardian, 20th April, 2016.
The heart's digital twin, The Economist, March 1st, 2020.
Alice's Adventure in Wonderland, Charles Lutwidge Dodgson, 1865.
Better data paves the way for improved accessibility, Financial Times, January 29th, 2020.
Automation finds home in building design, Financial Times, Eoin McSweeney, January 29th, 2020.
Artists and activists offer privacy hope as facial recognition spreads, Financial Times, Edwin Heathcote, January 29th, 2020.
Prospering in the pandemics: the top 100 companies, Financial Times, June 19th, 2020.
Moves to limit Big Tech still only half formed, Richard Waters, Financial Times, June 19th, 2020.
After disaster, The Economist, June 12th, 2020.
The destiny of density, The Economist, June 12th, 2020.
Which Towns and cities will be post-pandemic winners? Simon Kuper, Financial Times, June 25th, 2020.

Chapter 3
Info-telligence in the City

3.1 Info-telligence in the City

Big digital players, usually defined by a business model based on a big data platform that is able to aggregate and mine data, are positioned to profit from further value migrations also coming from the pandemics, if left unconstrained by local and global regulations that, however, all appear now more and more likely. For a start, they have certainly been greatly innovative and able to disrupt many status quo paradigms and capitalize on the dramatic surge of a knowledge economy where data and intelligence play a major role. These have also been creating the premises for a "winner-takes-it-all" competitive battlefield—a fierce arena of international and local competition where almost no one acts right now as a credible consumers' advocate. To this end, there have been calls from policy makers, independent thinkers, and association—not only they are advocating "digital dividends"—basically an idea to tax big digital players not just because of their tax optimization techniques (being digital, they usually structure their taxable income across the most advantageous jurisdictions—thus arbitraging existing rules on a great scale). But also, they have called for swift anti-trust action because of their huge value accumulations and market capitalizations (measured in Trillions of USD, in fact) that are suggesting some abnormal renting position. A digital, tax-based dividend would provide a straight transfer of money to ensure some funding back to cities and citizens. Or new rules and regulations could just create stronger property rights that can allow a greater and more explicit monetization of data for the people, when they cede information and—indirectly—intelligence on their behavior, inner feelings, and even their biometrical or genome-related biological setup. The design of data-cooperatives acting as unions to get greater bargaining power and distributing more fairly the value generated (And the winner is 2020) is an idea that has been discussed in the past, but that is all but easy to execute at scale. The development of new laws and of independent parties able to provide some fair price estimates on data, based on scientific, objective, and verifiable "white box" modeling, could also be useful, but

© The Author(s), under exclusive license to Springer Nature Switzerland AG 2021
C. Scardovi, *Sustainable Cities*, https://doi.org/10.1007/978-3-030-68438-9_3

hardly able to steer any structural change on how data and info-telligence are used and monetized. A growing and changing role for the financial services sector could, as discussed later on, instead prove a more practical and interesting option to consider.

Whatever the most effective solution is, it could be pursued to ensure that this emerging info-telligence is optimized and developed on a more transparent and fairer basis. It should then be designed and executed in the context of global cities. In fact, global cities are today able to attract the majority of the world population and will likely end up polarizing even more the sourcing of most relevant data and the following production of info-telligence. Data and info-telligence could then be traded for a short-term profit, but also with a view to sustain the maximization of the mid-term wealth being of the citizens—as the "smart" way of doing things become meta and fully sustainable. In truth, in a sustainable, global city, we will all get enmeshed in a distributed sensing-technology ecosystem. We will all then become more and more aware of our richness of data of the value of these, and we will then be able to trade info-telligence for money, in greater transparency, efficiency, and fairness. If this is happening to full potential, it could lead to a disruptive sociological and economic revolution, with a critical redistribution of control over information that will empower individuals, providing insights on their inner value and when, how, and for which end purpose to share them with others. Eventually, in sustainable cities, incentive-based, objective, transparent, and independent platforms could give people a small monetary compensation for the way they continuously create, almost unintentionally and often unconsciously, a continuous trove of data. These "efficient market" development would in turn support the design and creation of new, hybrid spaces at the intersection of the two dimensions. Cyber and physical will then get fused through a productive collision, with cy-phy companies that will be set free to successfully pursue not only their shareholders wealth creation (given the fair data monetization mechanisms set in place, cheaply and easily available) but also their stakeholders' mind and body well-being-fullness, as they will manage data via opportunistic sensing, crowd-sourcing, and ad hoc use case design and deployment.

3.2 The Economics of Info-telligence

The economy of info-telligence is an interesting one, and we would venture the main underpinning force that has led a number of companies to reach the trillion dollars valuation. A new set of "seven sisters" may be actually getting there, if we count, on top of the American "big digital" (Amazon, Apple, Google, and Facebook), the Chinese ones (Alibaba, Tencent, Baidu, and many others to come; see the recent performance of Pinduoduo). Data has been, in fact, compared with oil, and as such it needs to be refined to become useful—cleaned and tagged to fuel the AI/ML-powered engine that in turn, through multiple, energy-guzzling analysis, will be able to produce power (intelligence). However, as an old advertising motto was saying, power is nothing without control—e.g., intelligence needs to be put at good use, via solutions which are relevant and serviceable and able to address hard, relevant problems that really matter to individuals and firms. During all this "production process," from data gathering, refining, and "combustion" to produce intelligence, there is a huge value multiplier at play. Data can be sold as such or as "insights" (some embryonic form of info-telligence). On a pre-screened, more-tailored basis, they can be both monetized as online advertising or more traditional offline one and based on the personalized profiles of customers—this advertising market was worth around $ 178 Bn in 2018, according to Strategy&. Data brokers also sell personal information to banks and telecom operators, generating a further $ 21Bn revenues. These are very relevant numbers—but, as we may expect, it's just the tip of the iceberg to come.

In fact, it's on more proper intelligence that greatest wealth can be captured—Facebook and Google are hardly selling any data at all, but they do sell relevant intelligence about who is the best target for a given advertising and then for selling some specific stuff (Digital plurality 2020). Not to mention that you can use this intelligence to win an election and become the President of the United States. In short, data may become an asset class per se, as the World Economic Forum ventured back in 2011, and an entirely new, more transparent and fairer market may be needed, likely with some regulations attached. Similarly, to what happened in the battle between content and connectivity (aggregating platforms playing in between ended up winning the war), data owner and connectivity networks could both be relevant and fight a tough battle, but again ending up with a third party winning the war. Our assumption on this front is twofold. First, we would assume the winners will be the one able to competitively play on the creation of serviceable intelligence—e.g., defining and implementing relevant solutions that are able to crack hard problems and create new value in the economy. Second, we would assume that these designer and manager of intelligence (or info-telligence, as an even superior asset class, intangible and highly liquid) would be able to do it at the cyber and physical dimension level, acting consistently and, in parallel, seamlessly enmeshed way. In a post COVID-19 society and markets where boosted connectivity is leveraged to monitor individuals and corporates, similar solutions could make a world of difference—for security reasons as well.

A number of trends pointing in that direction are already evident. Amazon Web Services (AWS), the cloud-computing arm of Amazon and current leader in this new Klondike-like "over the cloud" run, has, for example, recently launched a marketplace that aims to make the trading of data as easy as possible, with buyers subscribing to feeds, agreeing to licensing conditions, and with AWS processing payments that are safely in place to ensure settlement, after the workings of an adjudication systems that is dynamically matching data supply with demand (as it would happen in an electronic stock market). In fact, this fluid information exchange and trading is not limited to formal, open air markets. In truth, the very existence of a black market for stolen information is also well documented. Information related to our credit cards, when stolen, is typically not used by those who were the digital thieves in the first place, as data is typically sold over illegal "wholesale" markets for payments data troves. It has even been observed that, in presence of large thefts of credit cards data, the price for this illegal data is moving down, as a consequence of the excess supply that is temporarily being created. In more extreme business models, the "big digital" platform owners can sell information to advertisers and political parties (or even dictators or terrorists—if left unchecked) and with highly sophisticated techniques to manipulate individuals and the public sphere, on top of potentially capturing (stealing) personal data on our digital (and physical) identity that can be valuable for manipulation or blackmail. Obviously, we all need to tread with data very carefully.

3.3 Data Are (Almost) Unique

Data can eventually be valued and traded as basically any other commodity (Data, data everywhere 2020), but with some caveats. In fact, data are not like any other "normal" and commonly traded good and service as they can be used time and again, almost like public goods. We say "almost" because some of the value they can generate for their owner is grounded in their being kept unknown and unusable to others (the data regarding the next, hostile purchasing offer in the stock market has an almost limitless value to the knowledgeable trader, as soon as he is the only one to know and trade on this information and able to enjoy an almost "risk-free arbitrage"). Also, some data may be more useful when managed via aggregated pools instead as a set of individual ones, for them to have greater statistical relevance and a greater chance to spot emerging patterns for predicting the future. Also, some data may be relevant for a few seconds (e.g., my GPS-based positioning, when travelling), while others can have a more lasting usefulness and serviceability (e.g., my music preference). They can have spillover effects as well, good and bad—good, if they, for example, may help to improve the healthcare of individuals, and bad, if they breach personal information to the detriment of the targeted people. Hence, the monetary value to the individual or business (or political) buyers can be quite different vis-à-vis their wealth-well-being value, if we consider and analyze data for their full potential impacts and spillovers on the economic and social ecosystems of the city. In fact, these spillover effects are mostly happening in the context of the city and of metropolitan, global cities specifically as cities—as emerging cyber-physical markets of the future—are also basing their competitiveness, success, and value creation opportunities not only on their ability to use data to become "smart," cyber, and meta, hence sustainable, but also on their ability to support the development of great info-telligence services and related businesses, to the benefit of everybody.

Hence, data are (almost) unique because the info-telligence economy they underpin can have dramatic consequences on the evolution and even survival of entire nations and even civilizations—consider, for example, the role data are having in the context of the fight of the coronavirus pandemic or that may have in a context of a major warfare or extreme climate change impacted context and for other extreme, catastrophic events. Data, as an asset class, and info-telligence, and as an emerging industry and potential framework on use to reinterpret the working of our economic system, should be managed with an inner and higher pursuit—shifting the emphasis from "how to make easy money in the smart city?" to "how to increase the overall value of a sustainable city?". A sustainable city is in fact one that is not just able to survive and protect the wealth and well-being of its citizens through extreme events, such as pandemics, wars, and extreme climate change scenarios—but also one that is democratic and with a good form of government, free and meritocratic but also socially sensitive and compassionate (S, social, and G, governance, of the ESG acronym).

To this regards, data and info-telligence do not have to lead to the creation of unchecked oligopolies and to the extreme polarization of its derived wealth, and it needs to ensure that the weakest counterparts in the society are also protected, included, and taken care of and with a fair chance to have a meaningful role in a cyber-physical reality of extremes—where almost everything is possible but where being left beyond and alone and marginalized by multiple digital divides is also a very concrete possibility—and often an unrecoverable one. It's not by chance that, as reported by *The Economist* (The Economist 2020), the number and the wealth of billionaires have increased sharply in the first 3 months after the coronavirus outbreak, led by Jeff Bezos, Bill Gates, Mark Zuckerberg, and a few others. To this regard, the emerging European Union "new digital strategy" and related regulations and policies are aiming to break the quasi monopolies of the likes of Google, Facebook, and Amazon, forcing them to share data as financial service companies are already forced to do (at least in principle, upon the sign-in of the client, following the PSD2 Directive). But even that may not be enough—it can help the sharing of data, but data are not the end of the info-telligence economy. According to *The Economist*, an "attention economy" should also be considered, for regulation and protection (Tristan Harris 2020) and, on top of this, as we suggested, an "intelligence economy" driven by AI/ML apps.

3.4 Attention and Intelligence Utilities

Cities can become "smart" if they can capture and somewhat manage data (obviously in a "smart" way). But they can evolve into cyber and meta if they can capture attention and create intelligence, linking the relevant dimensions of the digital and physical worlds. Sustainable cities can then thrive if they are able to develop utilities for them that can in turn leverage to support public and private value adding services and solutions. These are critical requirements for winning cy-phy companies as well. On the attention side, societal media platforms are now intermediating the way we construct shared truths and social relationships. And the winning ones are those that are better able to capture our attention. Network effects are then naturally promoting the emergence of dominant platforms that are then catching our attention more and more, while all other sub scale platforms are losing in relevance. This "winners take all" trend, in turn, is favoring the emergence of polarization, extremism, tribalism, and fake news that are able to influence most of the working of our cities—from political elections to commercial ventures and business practices, to friendships and romance, and so on. The EU GDPR (General Data Protection Regulation) has decisively addressed some of the privacy issues and has become the gold standard for over 120 countries globally. But it has not been able so far to address this "concentration of power" issue. This supply side power concentration is in fact just partially addressed by GDPR—it is difficult to see how this can really support us, when dealing with "big digital" media platforms. It is difficult to negotiate with Amazon (in ecommerce) or Google (on search engine), when they are asking us to sign-off so many (digital) pages of terms and conditions. These terms and conditions are of course non-negotiable and impossible to read and understand—unless you hire a GDPR experts army. You may, eventually, decide not to sign-off. But as most big digital information providers, given their quasi oligopolistic position, are requiring this, they leave no room to maneuver to us as a consumer. We cannot simply do that or we risk to become marginalized in a digital world that, because of its link to the physical one, would make us de facto big losers in both—also, have you tried not to sign-in to dominant players like Google or Yahoo! that are also acting as default browser on you smartphone?

Moreover, GDPR is mostly focused on an "ownership" principle when addressing personal information, and this bias leads to a straight, but also simplistic, yes or no on their access. Instead, it does not focus at all in building a more discretionary framework for designing "accessing rights" to information that could allow people to manage in a very simple way how these data are used and potentially monetized (i.e., potentially selling each and every data and with a specific use case as determining their price tag). Finally, it does not prescribe how policy makers could use such data on a cohort basis to pursue public good. Such pursuits are obviously possible and very relevant—they are making cities less smart and more cyber, meta in fact—to progressively support the city in becoming truly "sustainable," as this should be the ultimate goal of an info-telligence-based enlightened society. Without a clear set of rules and some policy making strategic directive, big digital (data cum info-telligence) platforms could capture at will our attention as citizen of "smart cities." They could distract and divide, or even outrage (or delight) us as voters, consumers and member of the fabric of society. Should this be the case, the advent of "smart cities" could turn into a major threat to democracies, if their central big data platform and related AI/ML capabilities are left unchecked, as the Cambridge Analytica scandal has shown too well. Hence, according to Tristan Harris, the European Union, and the other relevant policy makers, should create a new regulatory framework for large, dominant social and business platforms that have created vital public and private digital infrastructures that are now underpinning the first wave of development of "smart cities." Because of their role and relevance, these "attention utilities," according to Harris, should be required to operate in the public interest, with proper rules and licensing—and, we could argue, with required capital buffers, if they are the owner of large "cloud" facilities as well, not dissimilarly from what is required for banks and other financial intermediaries. These utilities, for Harris, would also be required to limit their information and intelligence extraction capabilities and message amplification, to make sure their influence (via microtargeting for advertising, for example) does not become too dominant—even more so as a protection to certain, weaker cohorts (e.g., children).

3.5 The Risk of Being "Big Digital"

Big digital players have come to appear as an almost unstoppable force in developed and developing economies across the globe. The top five of them, in the western world, have grown in market value by $ 2 Bn in 2019, then loosing much less than the market in the aftermath of the COVID-19 outbreak and then recovering fast to then break new records. In fact, four of them have consistently passed the "trillion" mark (Alphabet, Amazon, Apple, Microsoft—Facebook is still below) and have approached a total worth of almost $ 65.6 Tn in total, with an implicit expectation that they will be able to double profits in next 10 years (Big tech's $2Trn bull run 2020). What was looking as the unreasonable hype that is usually visible at the end of a long expansionary economic cycle, it is now appearing as a more and more resilient positioning, as the global economy is hit hard by pandemics and "big digital" companies are coming to play an even stronger part in our (new) everyday life. Not only they entered the crisis with a generation of $ 178 Bn of net cash flow in 2019 and with median sales growth of 17%—nothing suggesting a steady state, let alone an inflexion point. But also, they have shown how they have been able to keep their business and operating machines in place (they employ 1.2 Mn people just in the United States and invest $ 200 Bn a year). They have even come to play a role of "lifeline" utilities during protracted lockdown periods—Amazon first and foremost—ensuring deliveries of essential and non-essential goods, but also information services (Google) and offering new ways of social interaction at distance (Microsoft, Zoom, Facebook, and others). In fact, in areas that were deemed approaching saturation (e.g., Facebook's services, including Instagram, Messenger, and WhatsApp—they were reaching 2.3 Bn people at the end of 2019—out of 3 Bn of current active users of smartphones), the "social distancing" have been offering them an extra opportunity to grow (e.g., during the initial weeks of lockdown, social messaging have been growing by over 50%). Also, further growth opportunities have become apparent if reconsidered at the light of the development of the "smart city." Eventually, they have an even larger role to play in "sustainable cities," where the info-telligence play should get extended to all dimensions of social and economic life and centered around the individual and with the explicit aim to make it more resilient through crisis.

Indeed, they have become a critical actor in supporting the cities' future evolution—something should be done, ideally with their help, in pursuit of the common good and in democratic, liberal, and non-discriminatory ways. In the "smart city" model (or phase of cities' evolution), they can support local communities on general utility functions such as mobility, safety, energy consumption, and the likes. As they then develop towards the cyber and meta ("cy-phy") phases, by capturing and managing data that are more and more related to individuals, they also get an increased chance to control and manage things like mental health and stability, social isolation and crowd preferences, fake news and accepted truths, wealth polarization, and the emergence of un-democratic forms of populism or newly formed racism. As a consequence, "big digital" are facing increasing regulatory risks, which may handicap their ability to catch up on future challenges of an operational and business nature—they also face the continuous citizens' screening, on how they trace and account for fake news, or illiberal and violence-enticing views and opinions. For big digitals, the tech-lash (technology-backlash) (Big tech's $2Trn bull run 2020) risk means a number of things and all very relevant indeed. Fines are already punishing them for tax, privacy, and competition misconducts, but their sheer size still looks as almost irrelevant given their market cap (a mere 1%, as calculated by *The Economist*). More importantly, strategic regulations (with GDPR—General Data Protection Regulation—pointing the way and potentially becoming a global standard, followed suit by California's CPA, Consumer Protection Act) could curb their path to further growth and make them more bureaucratic and less innovative. Given their size and potential influence, they are also becoming the obvious target of policy makers of all kind that look to either curb or leverage their influence on global voters.

Further EU regulations, on open data and related AI applications for big digital (The Brussels effect 2020), are then aiming to create a "single European data space" in which digital information flows freely and securely—for the benefit of citizens as consumers of info-telligence and not as just owners of personal data. The open data movement, for example, is based on the idea that we "own" the data we produce and everybody should be incentivized to make a fair share of their data available for the public good and in order to optimize the social wealth that can be created from them in a sustainable and inclusive way. In the way to the open source movement, the democratization of data, if not leading to a democratization of intelligence, should at least lead to a greater level playing field on the subject of AI ideation, creation, and utilization—hence for the overall market of "info-telligence." Data sharing and crunching will then need, of course, to respect regulations such as the GDPR and the CPA, but this could be done via encryption and by analyzing data by cohorts, to address issues that are still relevant at the individual level ("homomorphic encryption"). For example, Oasis Labs, a start-up in San Francisco, allows the donation of genetic data to be used on research projects and with a full privacy on the donors identify. And "data trusts" could also be used as the basis to manage protection on privacy but also allow a fairer and safer data exchange—on the basis of a "digital services act." The idea of creating a "digital services act," imposing stricter rules on info-telligence economics, could limit their dominant position, but could also hamper their potential and critical contribution to the build-up of sustainable cities. While the main aim of these regulations may be the promotion of a more inclusive and fairer use of technology for the preservation of democracy, freedom, and compassion in a closely intertwined cyber-physical ecosystem, the net effect could be the setup of a bureaucratic regime that is seriously limiting the competitiveness of the countries adopting them, with their cities falling short on their targets of sustainability because of these. Without "private" big digital players, it would also be difficult to see how huge investments in info-telligence infrastructures and innovation could be financed and operated in a competitive way.

3.6 A "Physical" Missing Point

Big digital companies are already facing a renewed push by the European Union to break up or limit their quasi-monopolistic positioning in certain areas of offering. The EU is actually pushing the likes of Google, Facebook, and Amazon (Espinoza et al. 2020) to open up their troves of online data as financial services already have had to. As a potential counter move, these same companies could try to evolve and develop their operational and business positioning by extending their reach on the (data related) physical world. In fact, to play their chances at best and become the winning "cy-phy" companies of the future, they will need to address a few critical points. Their first, most obvious weakness is in fact their still limited access to data that are generated in the physical world through individuals' behaviors and within the context of complex ecosystems. While Google and Facebook are dominating in the online advertising world and, respectively, in the search engine and social network solution selling, nothing similar can be found, at scale, in the physical world. And while Amazon is in turn dominant in the ecommerce world and across most sectors of online retailing, nothing of that magnitude can be found offline. Hence the uneasiness of the EU on their concentration of power in the online world and (likely) more relaxed attitude on their still very marginal physical presence. While likely being constrained in expanding further in the digital realm, they could then try to grow in the physical one, by rebalancing their overall presence and competitive positioning.

For the big digital players, the best chance to complete their data and info-telligence dominance could then rely on the capturing and retaining of the data flows that are stemming from the atoms realm—they could, for example, aim at developing a strong foothold in the design and management of smart cities. That's why Google is trying to get into real estate development and urban regeneration (with its fully owned Sidewalks) and Amazon is entering physical distribution starting from supermarkets chains (with its acquisition of Whole Foods). It would then not come as unexpected an acquisition by Facebook of a club of gyms or restaurant chain or collection of theaters and cinema, as they would all help in complementing its offline social data flows. Crucially, all big digitals are also trying to enter the payment and money management worlds, as these would help them in matching the social data and dimension of analysis with the financial (missing) leg. Most of these missing points can be addressed by big digital players trough cyber-physical cities, where they can collect data via IoT and store them into public and private data bases, likely running over the cloud and with multiple AI applications to mine intelligence. Big digital could end up working for central government policy makers, to help them in making their country more competitive, and with cities' ones, to help them in becoming more attractive for leisure and business. Different sets of data could then be managed with different rules of access and use, with multiple and diverse players acting on the same, dominant data platform, with different degrees of freedom and the requirement to give back some of the intelligence created. In a way, policy makers would erect borders in the mirrored, digital world, contradicting the principle of open data itself (Virtual nationalism 2020).

In turn, other, non-big-digital companies could have a chance to play a larger role and get to a winning dominant position, should they be able to develop, from the basis of their offline data capture and analysis capabilities, an online data dimension as well. This kind of competitiveness, built on the basis of the cyber and physical dimensions, would then be a way to nourish and sustain the production and distribution of superior info-telligence—monetized in those key markets of the future that we believe are mega-, sustainable cities. We have referred to these winning companies, in other books, as "synapsis-companies" or "singularity" ones. We could as well reinterpret such definitions at the light of this current cy-phy discussion that is also going to determine how a city could be sustainable and successful in the future. In fact, a "synapsis" company would be able to aggregate structured and unstructured information, both available in the digital and physical online and offline, worlds, and using sophisticated applied analytics (including AI, hence producing "singularities") to support a new "intermediation model." Following this model, where companies could act as connector of multiple stakeholders, would be driven by the design and delivery of solutions that are addressing critical and relevant issues, for the ecosystem of reference and to create new, sustainable wealth—sustainable over time and inclusive with respect to many stakeholders, also because of the overall trust they would put on this cy-phy system. The "singularities" produced would then impact all kind of digital and physical dimensions, acting in parallel on the twin and intertwined dimensions and drive their resulting economic value creation.

Real estate and infrastructures would then need to adapt, as connectivity networks and data base should as well. Building, for example, should be redesigned to support and facilitate our digital experiences (think of the "smart working" happening more and more often from home—you need a proper physical, temporary office space at your residential building, on top of 5G connectivity and 3D videoconferencing capabilities). And data bases should also be re-localized to ensure proximity and to allow almost real-time sharing of information—this would be quite relevant to support, for example, self-driving cars and Fourth Industrial Revolution production applications where, again, the digital and physical dimensions are asked to optimally coexist and develop to optimize output and allow the pursuit of innovation. In this context, whether as an incremental development of existing big digital or as a renaissance of more traditional, mainly physical organizations, a new kind of company should emerge as winning model and able to pursue a more balanced and sustainable development across the cy-phy dimensions and supporting citizens in reaching a well-being that is as much intellectual as physical. This "cy-phy" winning company could then act as a mutagen—an agent of mutation, by introducing new designs and artifacts for the meta city for it to better address its needs and pursue an evolutionary path that is creating a value that is sustainable and inclusive.

3.7 Death of the Smartphone

The info-telligence business and related economics have both seen, in the last decade, a first, huge surge in their value creation (and migration) capabilities—and mostly for the benefit of few, large platforms and high-tech operators that we have dubbed as "big digitals." They include search engines (Google), online advertising platforms (Google and Facebook), extended e-commerce (Amazon), and producers of hardware and software that are supporting and enhancing our ways to store and use data and then produce intelligence (Apple, Microsoft, and the same Amazon and Google as part of their cloud offerings). Most of them have become, in turn, trillion-dollar companies, with a few others to follow. Other, smaller unicorn companies active in these fields for some vertical slice of their digital platform or for specific use cases have then followed suit. Most of these info-telligence value creation (and migration) has been greatly accelerated and been largely based on the technological development and market adoption of a specific, ubiquitous if not notorious, existential artifacts of our contemporary way of living: the smartphone. Since the advent of the Apple I-Phone, all kind of smartphones have appeared to reach an estimated 3 Bn people and taking up a significant chunk of their time (it's not uncommon to spend 3–4 h a day staring at our smartphone, if you look at the statistics provided by our very same digital companion). In fact, they have taken up so many essential and non-essential functions and leisure or working-related use cases to almost become indispensable to us. Without them we feel lost and on many dimensions.

In short, they have become part of an almost post human condition and a core component of the "smart city" proposition, to make our life even more digital and driven by our small screen and hence more productive and fun (assuming this is the case, when we end up spending a good part of our working and free time as hypnotized by them). Smartphones have then become the portal to our digital world and our info-telligence navigator that get us to full speed when living in the context of a smart city. Through them, the smart city we live in is burgeoning and constantly unfolding, allowing us to be clearly geolocated in space and time (also for monitoring purposes) and to know our preferences, schedule, and consumer patterns. Hence, smartphones have been adding location-sensing capabilities by the day that allow us to perform multifaced activities and pursue parallel social and business interconnections—enjoining entertainment when travelling and communicating while doing so gathering information when training and accessing food and even romance and sex in the meantime. With the multiple evolutions of smartphones, we have progressively come to understand the complex and broader reality of the city—using these as our remote control on the transportation, weather, social, and business apps available. This is then leading to new ways to co-live and co-work and "swarm"—where "swarming" refers to groups of people that are digitally interconnected for any reason and start acting together, in a somewhat un-obviously coordinated way, e.g., with flexible groups of people connected by cyber media that act and converge suddenly on certain physical spaces ("flash mobs"), then driving viral propagation to larger and larger masses.

Still, at the top of their success, smartphones will be facing soon an existential threat, as new technological gadgets become available (e.g., bracelets, pins, lens, watches). They will improve the diffusion of the on- and offline information being captured, mixed, and melted together to produce info-telligence that will help cities move towards their next, meta stage of "sustainability." Certainly, they will still be able to retain some kind of "central hub" strategic positioning and to play, for some time, the control room role. But, as new IoT technology will become more diffuse, ubiquitous, and implantable in the human body, android-like, it will allow the emergence of truly "cy-phy" people and communities, able to swarm in many other and more extensive ways, sharing thoughts and feelings, in real time and body to body—that will make emailing look like some prehistorical relict of the bygone past. New-generation IoT are already there and set to challenge the hegemony of the smartphone. For example, Cochlear is offering implants that allow the deaf to perceive sounds—as an internal artifact working for us when our senses let us down. Other virtual prosthetics are also helping blind people, with cameras capturing information and giving inputs directly to the brain. Special lenses are then able to provide deep external analysis on anything we see, in real time. And digital teeth have been made able to monitor a person diet and his level of glucose. New diagnostic sensors able are even getting good in sensing the human biome, monitoring our evolving health without the need of external, impact-based, analysis of blood and pressure. Technology is however not enough, if a take-up in adoption is not ensuring that these tools are actually used and perfecting themselves, as the AI app that powers them inevitably needs huge trove of data to gain intelligence. If this may look like a long journey ahead for people to extensively accept their becoming android and hence continuously monitored and connected to some kind of big brother, think twice. The emergency of the COVID-19 global pandemic has in fact provided the opportunity for their complete rethinking and for an extended (and sometimes forced) acceleration in their adoption rate. The principality of Liechtenstein has, for example, piloted a program to fit its citizen with biometric bracelets able to send data on their vitality body metrics—including skin temperature, breathing rate, and heart rate back to Swiss laboratory for analysis, meant to assess their likelihood of having contracted the virus. They will be offered to the whole population soon (and made compulsory?) (Jones 2020). It is not unreasonable to expect that other, under the skin, devices will be implanted and maybe made compulsory in the near future, as ID and payments tools and for healthcare purposes.

3.8 Architectural Evolution and the Redesign of the City

These always-on, rechargeable via radio frequency, IoT-integrated implants are progressively creating many new interfaces with our physical anatomy and without requiring a screen, or a physical hardware that is—no matter how light—still weighting a bit and can easily get lost or fall or be stolen. As our individual self becomes, through this, even more cy-phy and android-like, it changes our "quantified us" paradigm. We can earn more and see from greater distances, and we can store more information and tap into more intelligence and so on. It also changes our "quantified city" paradigm, as we can live in an enhanced cy-phy city and with more extended and intensive ways. Our technological prosthesis are then becoming sentient and symbiotically intertwined with our consciousness—enhancing and augmenting our interactions with each other and with our communities of reference. When this journey will be completed, and more direct and functional interfaces between cyber and physical "twins" will be available, smartphones will become in turn irrelevant, and screens of all kinds and shapes will also become relict of a bygone era. Uber will be activated as we raise our hand to call a (not yet in sight) taxi; and Tinder and Grindr will allow to connect nearby people for anything from social to romantic encounters—they will be activated as we just snap our fingers and we raise our thumb up (if we like the link). Smart cities will then evolve with a new set of paradigms, more centered on the individual and pursuing a sustainable balance between what is digital and physical. As a consequence of this technological upheaval and "death of the smartphone," architecture will also need to rapidly evolve to better host us as almost-cyborgs. New architecture, more apt to host and address a cyber-physically extended environment, will in turn drive the redesign of global cities and of their urban regeneration programs. This could also mean that living structures will be developed to ensure the continuous flow of information and intelligence and the senseable interactions between their digital and physical realms. Buildings and streets and other public spaces would then become wholly cybernetic systems that can regulate and adapt themselves, evolve and learn, and mutate even dynamically through time and space (e.g., with pedestrian sidewalks changing in width during the day based on the overall footfall flow). Everything will eventually get fully optimized—as only a cybernetic systems powered by AI apps could do. At San Babila in Milan, for example, streets and sidewalks will be remade to change dynamically through lighting systems and barriers and become one- or two-ways (or just pedestrian areas) during the day or week as a function of the people and car traffic and observed behavior of both. Streets and buildings could share information as we walk, on commercial offers proposed on a marketing of one basis (for the shops nearby) or offering advice on the suggest route to get to the next transportation hub and more—adding multiple levels of sensorial experience as we go by.

Buildings as well will get embedded with more and more sophisticated communication and learning systems able to communicate with our wearable (or implanted) IoT. Even our flat will start acting as an intelligent agent that has the ability to learn from us and the way we live or would like to leave, when at home. Our home experience will then change drastically: from living in a house to living with a house (Ratti and Claudel 2018) and with architecture becoming a form of interface or middle-wear and with internal design acting almost a browser to help us navigate the cy-phy space. Furniture and design objects could enable spatial cybernetics and the interplay of the digital and physical realms. These user interfaces could also maximize, at the same time, the collection of the offline data generated by our behaviors and provide feedback based on their new info-telligence. The end game could well end up being an augmented, living architecture, working as the hardware that is ready to host cyber-physically augmented individuals, able to live and work across diffused spaces and bodies and leveraging info-telligence as they go. For example, a "smart" building would today target a reduction in the energy wasted to keep a certain temperature in empty rooms by knowing when they are kept unused, for how much and vice versa. A cy-phy, sustainable city would instead target a given temperature for people themselves, by irradiating them with a warm light and by monitoring them with motion tracking sensors—as a result, an individual thermal cloud would follow people and not focus on rooms. A similar thing could be done at global scale with the global energy grid, with sunny regions sending energy to dark ones (via a global superconductor of spike temperatures) (Ratti and Claudel 2018). This augmented, living architecture could then target not only personalized utilities, such as the energy saving ones, but also the very notion of a (personalized) beauty itself, meant as a critical driver of attractiveness of the place and with tech devices and IoT that will be able to extend and augment our perception and experience of it, e.g., adding a given perfume to support a certain view and usability and (perceived) state of mind, or changing lights and sounds accordingly, to define and sustain a targeted, valuable "living proposition," given the context and the way we feel right there.

3.9 Where Would the Intelligence Sit

In today's smart city world, the design, building, and management of the built environment are embodied in digital bites that are mostly created through a human act and with the help of that technological artifact, that is, the PC, the tablet, or a smartphone. Somebody clicks on a website or taps on a banner to request the activation of an information or a service that is deliberately and strictly controlled by some kind of big data platform. In tomorrow's sustainable city world, co-design, co-building, and the co-management of the built environment will more often than not be activated, in a collaborative way, by IoT (25 Bn by 2025, according to Ericsson) acting as sensors on the world on which they are embedded (and likely including human wearables and invisible implants) (Spreading out 2020). The destination of these bites would also change. While in the past, most bites would remain in the physical place where they have been created (e.g., in our PC, or smartphone—at least as a backup), today's emerging trend is for the migration of all of these towards the big computing factories operated by AWS (Amazon Web

Services), Microsoft Azure, Google Cloud, and the likes. Indeed, small to micro IoT will be designed and built to reduce their energy consumption and size, hence as unable to retain a memory of the data they have been able to capture—all sent to some space "over the cloud." In the cloud space, a most centralized approach could then prevail (it has indeed been the case so far), as data attracts more data ("data gravity"). On top of scale, heterogeneity will also matter, as different sets can be more profitably mined together (the greater the troves of data, the better the AI algorithms can be generated and the most effective the end use cases will be).

Cloud solutions can then offer to cities the use of their number-crunching and web-based applications, at scale and with a variable costing approach. This market is huge ($ 230 Bn globally in 2019 and expected to grow to $ 355 in 2022 according to Gartner) and dominated by a few suppliers, with the lion share going to the top three (Amazon Web Services, Microsoft Azure, and Google Cloud). Once data are on the cloud, business apps can then follow and be moved (or created directly) there and with greater applied analytics on offer for the end users to count on—this creates concentration risk, some latency, and high energy usage that is required to move data around. For example, AWS, the largest player in the cloud computing business, presents, as of today, a centralized model where all the data are collected and crunched in a few places—big data centers that are safely guarded and designed as fortress-like but with almost systemic consequences, should anything go wrong. SWIM.AI, a start-up, is on the other side offering analytics to optimize the traffic flow of the city by optimizing our mobility as digital twin citizen living in a data center. It pursues instead 5G "edge computing," where data are processed in real time and as close as possible to the place where they are collected. Edge-computing can enable smaller local centers and optimize dynamically their use to reduce latency—for self-driving cars, a few milliseconds of reduced latency can save lives (gamification is also a powerful, albeit less socially useful, use case for the deployment of 5G and edge computing). A new strategy—dabbed "multi-cloud"—is also on the rise in the "over the cloud" computing battlefield (Altocumulus 2020). Its basic idea is that to make the "over the cloud" solution more competitive and sustainable, companies and cities should use and leverage multiple vendors, using multiple providers, even dozens of them—to better address diverse needs, specific digital tasks, necessary redundancies (which were laid bare, across regions and geopolitical blocks, during the recent COVID-19 crisis), and specific compliance requests which may include data localization requirements (e.g., Google Cloud has closed an agreement with TIM, a leading telecom provider in Italy, to be able to offer Italian based cloud infrastructure premises, for clients unwilling to have their data travelling to Palo Alto, in California). As a consequence of this strategy, and maybe counterintuitively, technological infrastructure and applications get re-fragmented and partitioned.

Scale can then still be reached if scope and focus are defined in surgical and logically consistent ways, so has to have an even more sustainable and hyperefficient and effective "meta (or sustainable) cloud," serving cities, companies, and individuals in their quest for their new cyber-physical dimensions towards better working and living—a well-being-fulness that is now even monitored through specific accreditations and raking such as the "Well" one (while Leed certification refers to the CO_2 emissions of buildings, Well is focusing on health, safety, and human experience and well-being through design). In a multi-cloud setup, an overarching, unified platform—a cloud of clouds—is ultimately trying to pursue both scale and specialization—the best of both worlds. Hybrid cloud is then a further variation, combining a customer's own computers with the ones operated by the big three cloud operators (or else). Applications can then be connected between the company and the vendors and among the vendors themselves. On top of new players, such as VMWARE or Red Hat (recently acquired by IBM), that are trying to offer this multi-cloud set up as independent managers, also the big three cloud providers are trying to offer something similar, with Google offering a multi-cloud technology called Anthos, Amazon offering AWS Outposts, and Microsoft Azure Stack, which let companies build local clouds hybridly linked to the vendor' bigger one. In a world of cities targeting a greater sustainability, this kind of solutions could address the safety and resilience of its cyber dimension and then potentially encompass the physical one as well. In an ideal endgame, multi-cloud, competitive structures could allow cities and companies to shift computing workloads dynamically across providers, based on their real-time availability and relative convenience and as a way to build buffers and redundancies. On the physical side, similar dynamically switching capabilities could be designed to manage and optimize disruptions to the global supply chain, with standard protocols in place to ensure continuity in the event of major shocks—such as a major pandemics or extreme weather conditions—and a dynamic approach to optimize the allocation and sourcing of physical goods, in normal times.

3.10 Geopolitical Implications

No matter what the preferred cloud strategy is, a number of geopolitical implications need to be considered with regard to where the data and produced info-telligence will sit and who the owner of the cloud and of the related AI apps will be (and which detailed use case will be specifically allowed to be pursued and with what endgame in mind). To make things even more complex, the overall mean of connectivity used will also play a role, given perspective risks and the potential ways that this could be used to stole or change data as they move around, hence influencing the resulting info-telligence elaborated, for business, political, or warfare purposes. More specifically, the fifth generation in mobile technology, or 5G, is playing a critical role in this—with huge geopolitical implications that will shape not just the defense and commercial strategies of military and economic blocks but also the very same future of global cities. Just consider the recent tussle on the Chinese Huawei—currently the biggest (and likely best) maker of 5 G telecom gears. The company, that is pitching itself and internationally minded, pro market, and fully independent from any political influence by the Communist party governing the People's Republic of China, is obviously well positioned to get to dominate the global 5G market as well, with consequent worries and retaliatory actions played by US government and propagating to its historical allies (UK more recently). A technological battle that is starting from the telecom industry is now spilling over to international commerce and is getting full blown to geopolitics—impacting on global currency wars, trade barriers and taxation, and even alleged industrial espionage, cyberwarfare, and acts of terror and so on. Vocal politicians are calling then for the creation of "national, domestic champions" and western versus eastern "big telecoms"—with 5G becoming a national interest sector as much as steel, cars, vessels, and artillery making had become in the aftermath of World War I and building up towards World War II.

Alternative and more creative solutions are also emerging, including "fully virtualized mobile networks," built on cheap hardware and controlled by software in a manner similar to computing clouds (5Geopolitics 2020). Rakuten, for example, is a Japanese technology group that launched (on April 8, 2020) the first full-scale virtualized network and with the ambition to challenge the status quo of main, at scale players (like Huawei) that are addressing markets and linking customers with highly specialized, proprietary, dedicated, and reliable hardware. A virtualized network approach would instead work on a principle were information is digitized, broken down into small packets, and sent to end customers via a dynamic route that is defined in real time and based on the available, best connections to destination, where it is then recomposed according to pre-set rules and then ultimately delivered. The 5G looks, at first sight, almost as an industry per se, in between infrastructure and technology, driving huge investments in networks and not necessarily with an endgame-change strategy in mind that would deliver the required or even a "quantum" returns on the huge investments ahead.

Certainly, a 5G technology can deliver much faster connections, but only in optimal conditions—with a base-station antenna in sight. In many contexts, previous technologies such as 4G could be as much if not quicker (when 5G does not work in optimal conditions) and would just not require any incremental investment. Aside from gaming, that is said to have been the first successful use case of 5G, much of the extra speed potentially attainable looks a bit, if not overrated, at least unclear with respect to the use cases that would really prove its business case—not justifying the huge investments required to develop the network. In fact, this could not be the case in the context of sustainable cities, where the cyber and then meta competitive dimensions are reliant on an IoT system that works in almost real time—leveraging the computing power of network-based stations that are placed on the "edge," e.g., near or in the middle of the city itself. The endgame is then the perfect coordination between cyber and digital twins that could work, think, and act almost in parallel, with zero-time lapses, to guide self-driving cars or manage robots in factory floors. This would then strongly support the business case related to 5G, in terms of overall potential returns. Albeit, it would still be unclear for whom the value would be created and then captured. Not necessarily for the benefit of the 5G owners, if they would keep being reliant on pipelines of communication, with little control of the data and information being shared, let alone of the intelligence and related, critical use cases that could be created and exploited in the overall system. Certainly, being the designer and owner of a 5G network can allow multiple competitive advantages and an advantageous view point on the emerging trends of use and solutions generation—but success is far from being assured, with more responsibilities to come and the very close watch of regulators and policy makers that will ask to monitor how all this super-connectivity is being used.

In fact, in sustainable cities, IoT networks can become the critical infrastructure that determine the overall competitiveness and attractiveness of the place. If these are then managed by 5G in a more competitive way, with 5G getting dominated by a Chinese company, it follows that this sustainable cities strategy could become a source of geopolitical weakness for much of the western world. This would change if the virtual network approach, as promoted by Rakuten, has a chance to win or at least to develop into a viable alternative. In that case, provider and users would be able to switch from one base station to another and likely leveraging different suppliers and solutions and ideally ending up replicating all functions of a network in a software or patch of many software. This would require the promotion of an open architecture approach for the networks that will be running, more and more extensively, our cities—and potentially the build-up of a "telco cloud," able to use, leverage, and control multiple gears from multiple sources. Its investment cost should also be much lower and allow for the creation and management of some required redundancy, spread across multiple network suppliers—something we have learned to value much more after the COVID-19 outbreak. In this way, these diffused IoT networks could help in driving forward, and safely, the development of the sustainable cities to come, in an eco-friendlier way and reconverting our global stock of infrastructure and real estate, and overall built environment, in a seamlessly cyber-physical way. To do so, they will need the critical help of "over the cloud" big data platforms and of the communication technology that is required to keep their data blood flowing. And, more importantly, they will need the support of critical AI platforms able to mine billions of new data flowing per minute and generate millions of idea to test for relevance and correlation and for their incremental information value contributed to the search for truths that are relevant, serviceable, and useful.

References

And the winner is, *The Economist*, February 22, 2020
Digital plurality, *The Economist*, February 22, 2020
Data, data everywhere, *The Economist*, March 1, 2020
The Economist, June 27, 2020 issue
Tristan Harris, EU should regulate Facebook and Google as "attention utilities", *Financial Times*, March 1, 2020
Big tech's $2Trn bull run, The Economist, February 22nd, 2020
The Brussels effect, continued, *The Economist*, February 22, 2020
Javier Espinoza, Madhumita Murgia, Richard Waters, Big Tech will have to share data under EU proposals, *Financial Times*, February 19, 2020
Virtual nationalism, *The Economist*, February 22, 2020
Sam Jones, Liechtenstein rolls out radical COVID-19 bracelet programme, *Financial Times*, April 16, 2020

Carlo Ratti and Matthew Claudel, The city of tomorrow, Yale University Press, 2018
Spreading out, *The Economist*, February 22, 2020
Altocumulus, *The Economist*, March 12, 2020
5Geopolitics, *The Economist*, April 10, 2020

Chapter 4
Sustainable Cities and Climate Change

4.1 Getting Hot

Global metropolises are getting hotter and hotter. They are certainly doing so because, as we discussed in a previous work (Scardovi 2020), the competition among countries and even geopolitical blocks—down to the very local level—will be increasingly fought among cities that in turn will become bigger and bigger, and richer and richer, further promoting a polarization effect, between the wealth and well-being of the people joining them and the rest of the left behind. In fact, global cities are already competing in terms of business and leisure attractiveness (the B-leisure index, as developed by *The Economist*) and for the sustainable growth in wealth and well-being (our "wealth-being") they are targeting. Global and local cities are all pitching to billions of people to get the most and best of them. Not just current and prospective citizens, but also travelling businessman and tourists and, of course, investors in any real asset that may be associated with them. Beauty is also, obviously, a determinant of people's attention, preference, and end choice, and as driving factor of the decision-making process of perspective "customers" (meaning, people that may select to spend time and money on a given city vis-à-vis many others and for whatever reason), it could be analyzed not only in terms of the natural resources of the original, pre-build place. But also, because of the exclusive features of its architectural and infrastructural developments (including public, open spaces). And technology, as we have argued, is the next big thing to come, to drive the competitiveness and attractiveness of cities—from new developments to large regeneration programs, everything should get not just "smart," but "cyber" as well and actually "meta," to ensure the optimization of the interplay and suffusion of the cy-phy dimensions that will determine our future way of being.

Still, beauty is not enough if "livability" is also not assured, with lots of clean air, a bearable temperature, and humidity across the four seasons of the year and a limited (and well-covered) risk of extreme events, such as hurricanes or earthquakes. All these factors are usually captured in a static way: Dubai is of course too hot to

© The Author(s), under exclusive license to Springer Nature Switzerland AG 2021

C. Scardovi, *Sustainable Cities*, https://doi.org/10.1007/978-3-030-68438-9_4

live in during the summer, but it's great in the winter season, and Singapore is quite humid for most of the year, but it can be managed. And Miami is great—just leave the city during the occasional typhoon—along with Los Angeles the chances of a getting a "bit one" (hearth quake, for Los Angeles) are anyway fairly low. Or maybe not, if this climate change thing is much more than the usual fad, pitched by consultants and by the occasional scaremongers from new populist parties to win votes? In fact, all is truly changing and fast and (this is the worry), in a highly un-linear and hence potentially, at some point, uncontrollable way. Take a look at the numbers. Global average temperatures have increased by more than 1 °C since preindustrial time, but current forecasts are showing best scenarios heading towards an increase to 2° and worst ones to 3-4-5 or more, with disastrous effects and implications. The Paris Agreement long-term temperature goal to keep the increase in global average temperature to well below 2 °C above preindustrial levels, and to pursue efforts (but with no commitment on results) to limit the increase to 1.5 °C, was negotiated and agreed by 196 member states in 2016. It fell short already in 2017, when the newly elected President of the United States Donald Trump announced his intention to withdraw the United States (one of the major polluter) from the agreement. Even if the United States should reconsider this, the Paris Accord target is widely deemed as a challenging one and unlikely to be met, absent major discontinuities (COVID-19, with its huge global lockdown, has almost paradoxically helped a bit, but in a very temporary way—the real risk is actually that, aside the short-lived reduction in NO_2 and CO_2 emissions driven by the lockdown, the sense of urgency and priority that climate change had just gained at the end of 2019 will vanish due to the health and then ensuing economic crisis). All in all, the world is not on-track to meet its agreed target of limiting warming to 2 °C, and it's not clear what could change its course. Under current policies, expected warming will be in the range of 3.1 °C–3.7 °C—as shown in Table 4.1. And the IPCC report, released in October 2018, states that "drastic action must be taken in the next decade to avoid catastrophic impacts." These would be hard to recover if these un-linear developments are producing a snowball effect, with limited options left to stop the avalanche or secure a "plan B"—leaving aside the Elon Musk project to colonize Mars as a backup, should also the development of Tesla electric cars not be enough.

4.2 Sustainable Cities

The World Health Organization (WHO) has recently estimated how 75% circa of humans might be city dwellers by 2050 (a similar figure, 70%, is expected, according to the United Nations), confirming the biggest and fastest trend of shift to urbanization in human history and driven by China as posterchild. Certainly, as discussed, the global pandemics effect could impact on this numbers, even changing their trend in the short term, with people moving back to rural areas—still likely just off large urban areas. But it is very likely that the secular trend would be confirmed and regain strength in the mid-long term. No matter what the final number of new urbanites will

Table 4.1 The world is not on-track to meet its agreed target

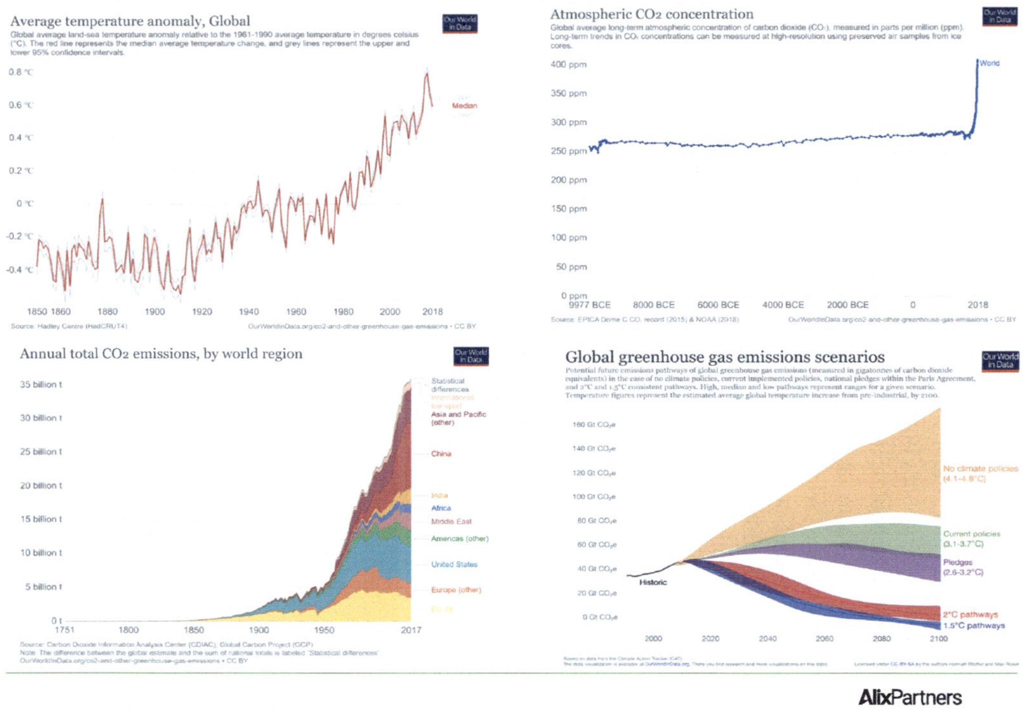

Under current policies, expected warming will be in the range 3.1 °C–3.7 °C

be, this migration, and its impacts, will be huge (a good target should be a further 2–3 Bn people moving from rural areas to cities in around 20–30 years). These new smart city inhabitants (along with the existing residents) will have to adapt to urban areas that are particularly exposed to natural disasters and climate change (Green 2020)—something that has also been building up in a secular way and driven by the unprecedented rise in carbon emissions. In truth, we are already living in a climate change emergency state right now, with things that are most likely to get worst before getting better. On this basis and actually because of climate change, a further set of large migrations could be expected, and not necessarily from rural to urban areas, but also within rural and within urban areas. Let's discuss a few examples to consider the potential consequences on our future geographical footprint—this will also influence the global, geopolitical competition and the success of the cities we live in today, with huge value creations and migrations within real estate and infrastructure as asset classes. What about if, for example, the average temperature should increase by 4 °C and not by 2? Would Milan become the new Dallas of Europe (not for its oil, but for its hot-like-hell temperatures) by 2050? And what about Riyadh that, in the middle of the desert, would already be unlivable by now, if its buildings were not kept refrigerated and de-humidified most of time and thanks to the use of oil-derived (and CO_2-producing) energy? Would the livability of successful global metropolis like Dubai and Singapore, New York or Miami, and Los Angeles change significantly?

It would, of course, and we would expect large, almost unprecedented (since the glacial age) migrations of people (and value) towards the antipodes of the earth. Expect the real estate price of Greenland to go up, including for seaside vacation second homes (would this explain the decisions of Donald Trump, a seasoned real estate investor, to go short on climate change—by dropping out of the Paris Accord, and long on Greenland—by pitching to Denmark his intention to buy the all lot?). And expect luxury villas on tropical islands like Seychelles or Maldives near the equator to get quite cheap (they would become literally unlivable, if not wearing an astronaut suite), assuming they are not already underwater by then (something already predicted for Maldives to happen in the near future). More importantly, environmental, social, and economic effects will also be very relevant, if not utterly dramatic, even assuming a + 2 degrees scenario. The problem of these kind of targets is in fact their relation to the "mean," with little attention paid to the "standard deviation" of the normal (Gaussian) distribution of frequency that is describing the expected weather conditions behavior through time (assuming we can rely on a normal distribution and not on something else). What is the relevance of keeping it to a 2° average, if the temperature peaks will reach 45° or 50° for a couple of days during the year and with a 99% probability and with resulting wildfires and death by heat, as recently observed in New South Wales in Australia (with 25.5 Mn acres burned in a few week)? It does not really matter how is the life during the remaining 99 days if you are getting killed or your house burn to ashes at least 3 days in a year. And what about the impact on extreme events such as hurricanes or landslides? It doesn't really matter if they happen once or two times over a 30-year timeframe, if those events are destroying half of the buildings, the relevant and critical infrastructure, and killing thousands of people and animals in the process. Livability is in fact defined by standard deviations as much as by averages and cities are becoming not just hotter and hotter, but also unpredictably so—because of extreme weather conditions—they are becoming more of a norm than an exception, forcing a complete rethinking of the safety standards that should be considered, by the rule of law and by developers and investors, when building a property. That's why, most proactive cities (Milan as a best practice among many) are already reviewing the safety standards and minimum acceptable requirements for new developments and old real estate stocks, for example, asking impermeabilization solutions and measures that could face a snowstorm of half a meter or stronger and bombshells of water that could precipitate in less than half an hour.

This increasingly hot, extreme climate event-prone trend could potentially put certain cities out of the people-attraction game in a few years, forcing many others to radically reconsider their approach to regeneration, new developments, and even full restructuring of the current stock of real estate and infrastructure if they want to remain safe and competitive and cyber and physical, hence sustainable—without just pretending, poker face-like, that they are "smart." This will likely require a proactive transformation approach to the design and management of cities. Designing, building, regenerating, and converting current metropolis into "sustainable cities" should become the main target of this and of the next few generations of policy makers and of finance, real estate, and technology key decision-makers from the private space. Investing in the livability of our cities should not just allow great opportunities for financial returns, but also act as a prerequisite to defend, preserve, and enhance the value and attractiveness of the main territories and industrial sectors of reference. It should also come as a prerequisite to preserve the \$ 200+ Tn worth of stock of residential, commercial, and mixed-use real estate asset class, the largest investable asset class by far at global level. This "re-tooling" of current cities into sustainable ones should then form the basis to tackle climate change and ensure the sustainability of the earth ecosystem—our only one, for the time being—for many years to come. Cities are not just one of the most exciting and vibrant markets to try to pursuit and address—as smart and cyber, meta and cy-phy. They also stand as the most promising way to ensure the continuity of our ecosystem and of mankind, with the help of finance, real estate, and technology—and a bit of creativity—all contributing towards their sustainability.

4.3 It's All About Carbon (Dioxide)?

The climate change potential impact on the economy is a scorching topic for discussion and is becoming a red-hot one for equity and debt investors alike. They are now forced to take this seriously by policy makers and regulators, as well as by media and end consumers that can affect companies' reputation by choosing with their feet and hence determine their commercial viability. Investors can also decide whether to invest on a publicly traded company or a privately held one on the basis of its "greenery"—as it ticks the box of ESG and does contribute to the transition towards a safer more stable environment. As a result, the top managers of all kind of industrial sector are now rapidly changing their attitude towards the "green" issue: not something good for a speech or a week-end relationship-building event, but a key hurdle and constraint to consider when making any kind of strategic decision across their company's value chain that can generate CO_2. Climate change impact is in fact mostly driven by carbon dioxide emissions. With CO_2 concentration levels in the atmosphere that are now well over 400 ppm (actually breaching the 420 parts per Mn for the first time[1]—the highest levels in over 800,000 years of human history—we are now globally emitting over 36 billion tons of CO_2 per year, and this number keeps increasing, almost unstoppably). A large amount of CO_2 is then embedded in traded goods—this means that some countries' and cities' emissions increase, while others decrease when we look at emissions based on consumption rather than production. There are large inequalities in CO_2 emissions as well, with the world's poorest countries that have contribute less than 1% of emissions but will be the most vulnerable to the worst impacts of climate change. This equation will also be polarizing between big cities and rural areas but could not be necessarily so for cities which are post-industrial, just "smart," or truly sustainable.

Given the stock of CO_2 and its progression in future years, climate change developments and potential catastrophic impacts will be highly complex affairs and most likely nonlinear. This makes any kind of planning and estimate of future trends and expected outcomes an almost impossible task for traditional statistical conjoint analysis to tackle—posing significant issues to public institutions, but also to lenders and investors active in real estate and infrastructure for global cities. Jakarta, for example, built as it is on a swampland, is sinking so quickly that the government is relocating to a new capital. Floods wreak havoc almost every month in Bangkok, Dar es Salaam, and Nairobi. Even Sidney is exposed to the hot-climate generated wildfires, which are obfuscating the horizon for weeks and mixing dangerous particles in the air its citizens breathe. Not to mention the many cities and places that will be submerged by the oceans, as ice lands melt and retreat, and they are hit by extreme and lethal heat waves in the meantime. Seismic disasters also loom high, with the likes of Tehran, Kathmandu, and Istanbul built on geological fault lines. And mudslides have also become a too common event for many places in Italy. Even without considering the unfolding of dramatic, media-catching stress events, current situation is already silently harmful and honestly unsustainable, with thousands of earlier deaths going almost unnoticed. Most developing cities have become pollution hotspots. Simply breathing the New Delhi air on a smoggy day is equivalent to smoking 25 cigarettes.[2] We may hope, but cannot reasonably pretend, that the situation in Milan or London and Berlin or Paris is materially different from that. For all the talk of the smart cities, and (our suggested) development into cybernetic, meta places, able to join together the dots of the cyber and physical

[1]Nato, Vostok ice core data, JR Petit et al.

[2]http://maps.who.int/airpollution/

dimensions, the thorny issue and most pressing and apparently uncrackable problem for the future of the cities is how to ensure they stay safe, recovering healthy and hence sustainable from a wealth and well-being perspective. Smart (or cyber/meta) cities and their stakeholders, namely, public institutions, financial intermediaries and banks, real estate developers and investors, and telecom and other technology providers, have a large role to play on this. They are also increasingly called by policy makers, regulators, and institutional investors to act consistently. Because it's not just about carbon dioxide and the way to reduce it—as it is all about managing climate change risks (and potential returns as well) and governing their ineliminable trade-offs, in an increasingly complex society and challenged markets, before we reach the tipping point and before any "second sleep."

4.4 An Anthropological Phenomenon

Carbon dioxide and climate change are not built or happening in a vacuum, and they are not exogenous in nature—it's not like the notorious comet hitting the heart and leading dinosaurs to death (according to one of the most popular theories explaining the disappearance of dinosaurs). It is, nonetheless, a potential terminal event that will require all of our best resources to understand it better, predict, manage, and survive. For the sake of this challenge, we may then reasonably assume that climate change is humankind induced and therefore anthropological in nature. Whether related to our behavior as individuals, private companies or public institutions and acting in an organized or mostly uncoordinated way, it all starts from the way we work and live, consume and invest, produce and save, or waste and recycle. And vast troves of data, apparently uncorrelated to the main question at hand may be helpful to generate insights—as eventually everything is related, in a space-time framework, even if their nexus could be unobvious, and almost impossible to see and detect for the human eye. If we believe climate change is anthropologically based, then cause-effects relationships need to be found and then managed, at the level of our behavior, and include, in their investigation scope, individual lifestyles as well as companies' strategies and public policies and the overall context of regulations and the rule of law. In fact, humankind behavior can be associated to climate change. This, at individual, company, and institution level, is influencing CO_2 emission and could be analyzed in a cross-section way (i.e., running analysis with the help of artificial intelligence, in real time and across different ecosystems and looking at the best ideas that could explain, influence, and curb carbon emission), given limited data history availability and relevance.

As anthropological phenomenon, it can be influenced in a number of ways, including public policies and taxes, education, and trading. But private, free markets could have an even greater role in this. Among the key industries that are thought to be in a position to influence decisively the investment-consumption cycles of individuals, companies, and institutions, the financial services one is at the center of much debate and speculation. It could support optimal global reallocation, to favor a smooth and sustainable transition to a greener economy through their lending policies and investment management decisions—commercial banks could punish big polluters and support green-savvy companies; and asset manager could decide to sell or buy stocks on the basis of their relative contribution to the environment, among many other things. Eventually, banks and investment managers could be forced by regulators to reconsider their allocation policies—with greater capital charges or reliefs, based on the CO_2 emissions attributed to their borrowers and invested companies. Not only that. Climate change, as critical discontinuity in the history of humankind, with impacts of likely Biblical proportions, will influence the quality of the financial institutions' assets and earnings in a big way. In fact, financial services could make two kinds of potential mistakes, when considering lending or investing in a counterpart of asset.

The first "type 1" error could then be defined as giving money to borrowers (or assets and, more broadly, to cities) that are heavy polluters, hence doing bad for the environment. Big data, AI, and the produced info-telligence could then be used to steer lending and investing policies away from those bad counterparts. Financial services would pursue and support the "Good" for the environment, but not necessarily in a profitable way. It is unclear if this "doing Good" could be made sustainable by means of "softer" regulations and fiscal incentives alone, as it could be a costly and risky journey on a very bumpy road. There is however a "type 2" error as well that is by no means not less relevant than the first one. This could be defined as giving money to the wrong counterparts (company or asset) that will be, business-wise, negatively impacted by climate change and hence unlikely to perform as expected and unable to repay the loan (if the money is debt) or return the capital (if investment) and no matter how much "Good" they are doing for the environment. It could be the case of a wind-farm in the middle of the ocean—it could contribute to the shift towards greener forms of energy, but also run an expensive setup and with a market price that is not allowing the company to pay back its loans and preserve its equity capital. More specifically, this "type 2" error (giving money to counterparts or assets that are not doing well for themselves) could then be further differentiated (and following the framework introduced by the Basle Committee and the Bank for International Settlements, or BIS) into a "physical risk" and a "transition" one. The "physical risk" refers to the chance that some companies will experience stranded assets and physical damage because of climate change. Think of any disruption that could happen to a manufacturing plant or distribution network—including their inability to work for a few days, because of floods, typhoons, droughts, and so on. Or of any damage that could also affect people's health, for heavy pollution or heat waves, that could then be unable to work or consume. The "transition risk" refers instead to the chance that some companies will experience uneconomic and unprofitable business models because of climate change—as they are unable to address the transition to a greener economy or are just missing the right timing, doing this too fast or too late. Think, for example, of the wind farm example above: maybe the right strategy, but too early. Or of the automotive companies still doing most of their cars with an old diesel-based combustion engine.

4.5 The Risk for the City

Climate change risks, of a type 1 or 2 error, and with reference to physical and transition risks are very much relevant for lenders and investors. They could be "punished" by policy makers, regulators, and (more importantly) normal citizens, if they go ahead and give money to companies or assets doing "Bad" for the environment (type 1 error). Consumers, specifically, could just vote with their feet and walk away from such companies and assets (real estate included). They could also experience severe losses, if they lend to troubled companies or stranded assets, or ensure climate change related risks, or invest in companies underperforming the market because of climate change. Even more importantly, at the light of our discussion on the future role and perspectives of cities, climate change risks are critical in determining their success or failure and in a number of ways. For a start, a city, or neighborhood and down to the building and flat level, could jeopardize its future if it keeps doing "Bad" for the environment. It may not suffer directly as a consequence (its heavy contribution to pollution reverberates across a much larger area), but it could become subject to fines by regulators, curbs, and constraints to future development by policy makers and—if banks and investors are going for the "Good" thing, they could see a short stop to the financing opportunities. Their debt will become costlier, and little equity will also be available (if not at a much higher target IRR, as offered by a few "bad and ugly" speculators). More importantly, citizens (people and companies) could just vote with their feet. Why establish their residence or main headquarters and offices in a heavily polluted city center? Given the increasing opportunities of smart working from much nicer and cleaner places, the competitive pay on offer for the same job should be inversely correlated to the CO_2 and NO_2 level of emissions in the atmosphere—with their associated impact on health (statistically sound analysis have shown their brutal impact on people life's expectancy). A bio-value proposition, of total wellness, green cities that are circular economy based should become a more-catchy slogan than the old, industrial heavy, steel, and carbon-based cities of the post-World War II bygone era. The pandemics should, just in case, make all of us just more aware of that.

But the exposure of cities to climate change related type 2 errors could be even more important and with short-term impacts. On one side, cities are obviously subject to physical risks. A number of global metropolis, with tens of millions of people living in them, will be—literally—underwater in a few years (Jakarta is an obvious example, with around 36 Mn people and a plan to relocate the capital of Indonesia towards a higher place). Others will suffer droughts and floods with an increasing frequency and severity. Others will just become scorchingly hot and unbearably humid—livable just indoors and with huge energy costs for ensuring the required climate conditioning. Other, less eye-catching trends are also relevant, such as the cost for infrastructure maintenance, repairing, and upgrading, also driven by more extreme weather conditions. The insurance costs for certain, extreme weather-related events have then already spiked, and certain tail events (less tail by the day) have just become uninsurable. In short, no matter what new designing and building technology could be employed, the expected costs from increased physical risks will become more and more relevant, making certain cities less economically viable. If physical risk will impact heavily on the expected losses, the transition one will brute-force radical value migrations across cities, regions, and even entire countries—maybe unprecedented in the planet history (leaving aside a few glacial eras). The attractiveness of certain places will change significantly, with Iceland (for example) benefiting from the ice meltdown and maybe becoming a tourists' target location for summer holidays and seaside living, and other places just getting fast out of fashion, or forced to reinvent themselves (artificial snow could save some winter-sky locations, specifically if new technology could create artificial snow that is more resistant to mild temperatures as well).

If we think metropolitan cities are explaining a large part of the global GDP and that real estate is, as an asset class, the largest by far at global level (larger than the sum of all equities and bonds), it follows that considerations from climate change-related risks (and returns as, as made obvious in our examples, there will be ample opportunities as well—with new value to be created and a vast amount being migrated among different cities, regions, and countries) should be central to any kind of global asset allocation consideration or investment and development decision—that is going to bet on a city instead of another or developing a new one instead, to migrate an old one getting underwater (as it is happening in Indonesia). All kind of decisions related to real estate development or refurbishment, urban and rural regeneration, and infrastructure investing (including mobility infrastructure, but also digital connectivity and telecommunication one) should investigate climate change future scenarios and understand how these could correlate with risks and opportunities and the future value destructions, creations, and migrations to come. How about planning a real estate lending portfolio across Europe without having a clue on how climate change will impact the physical asset and the attractiveness of its location? But also, how about lending to agribusinesses in Asia without having a fair estimate of the changing conditions, yields, or even feasibility at all of having crops and breeding there? And how about planning huge infrastructure spending splurges, to support the recovery of the global economy after the pandemics, without having a clue of how people and companies will change their behavior because of that expected change of biblical proportion that is going to be climate change—likely tens of times as impactful as the coronavirus? If we then agree on the relevance of these kinds of questions, just a small issue remains: how to make predictions on one of the most un-linear, complex, and unprecedented phenomenon like this?

4.6 AI and the Search for Serviceable Truths

The world is becoming two-dimensional, we could provocatively say. On one side, the physical real is (and will be) still relevant, and understanding its key variations due to climate change is becoming an almost ontological question for sustainable cities to answer. On the other side, its cyber dimension is growing by the minute in scale and relevance. The amount of data captured and stored in a week is likely, in the near future, going to match the one accumulated since the start of the mankind and with an exponential trend. A lot has then been discussed regarding the ownership and protection of data, of a personal or anyway private nature—the debate has become so controversial to get the full attention of regulators, policy makers, and media. This is very understandable, but it has also led to forget about the incredible richness and deepness of the data that are publicly available and "open source" (i.e., they can be easily linked and aggregated via API that can source and draw troves of data that are already stored and that keeps being captured by the minute). They are not only huge and growing exponentially (and available for free) and with a trend that would see, in our view, a greater availability of open source data, because of regulatory and policy maker pressure, but also as a democratic ask of people around the world. They are also well representing both the structured and unstructured dimension of data: from statistical information on macro and microeconomic indicators, including financial statements and trade accounts or analysis on "ease of business" or "corruption indexes," to social information regarding whims and preferences of companies and people, as they change in real time, at the speed of a TikTok video. If we believe that climate change is an anthropological phenomenon, then it needs to be investigated from the people and companies' behaviors and up. For this, publicly available, open source data are going to be a very rich resource to use, which can of course improved if other private data sets are then layered on top, but no time should be wasted to start interrogating the data and AI, across time and space as they are all interrelated, to ask them what we would never get to know if counting on our human brain.

If data availability is not the issue, how to proceed then to derive intelligence from these, on something that has never happened before (leaving aside the glacial ages) and is potentially driven by so many causes, with complex and un-linear relationships to make any traditional, conjoint analysis approach very difficult and likely irrelevant? With so many potential explanatory variables—from what we eat and what we wear and how we travel to millions of other, all highly correlated, potential independent variables that could bring each some extra information contribution—there is also an issue of "where to start from?" and "what to leave behind?". Any decision in that sense would take out some "information power" on one side and, on the other, would introduce a "cognitive bias" in the process as we, as humankind, are prone to bias in the way we pose questions (usually, we go for the ones for which we feel we have an answer) and we make hypothesis (also, what we feel—sometimes, irrationally—that could make more sense to get to that answer). Also, while using AI (artificial intelligence tools) would help on those fronts, the alternative of relying on "black box" approaches could make the cure worse than the illness. In short, we need to find an approach that could mine troves of data by the minutes and generate independently tens of millions of hypothesis for testing in a few minutes and in an iterative way (as we may choose to aggregate some of the thousands more relevant into a lower number of correlation drivers). With no human bias in between, but also with a "white box" that is able to show "what is driving what and for how much," to get then to the "because" and the "hence" that would then lead to potential solutions—likely then harnessing all the global knowledge, including research papers, patents, methodologies, and crowd funding intellectual contributions that is potentially available—on an open source basis. Information and intelligence, and even "truth," maybe not absolute but still objective enough to be relied upon, is still of little value if it cannot be acted upon, to make the two intertwined dimensions of the city (the physical and the cyber) more efficient and effective and more productive and sustainable—and driven by the pursuit of making the world a better place for humankind to live in. Serviceable truths could then be facilitated by such an AI platform and approach, but would then need a lot of human passion, emotion, and physical and intellectual energy to also become "serviceable."

4.7 It's All About (Climate) Risk Management

If we associate technological innovation with the extensive use of natural resources in a post-industrial world, what is happening with climate change looks like the revenge of the physical dimension over the digital one (even the use of AI algorithms produces a lot of CO_2 emission, because of their intensive energy usage—not to mention the block chains used to mine Bitcoins). Depauperating scarce resources (we are learning the hard way) is a potential cause for extinction, as we grow our riches out of the impoverishment of the environment and forgetting, in the process, to respect the basic rules of nature. Following a popular interpretation (albeit with little scientific evidence), we could also attribute the emergence of more frequent pandemic diseases (like the recent COVID-19 one) and of more antibiotics-resistant viruses to the overall imbalance produced by our predatory economic growth models on our ecosystem and only livable planet of reference, from an ecological and biological point of views. We have taken for granted the rewards coming from development and exploitation, but scarcely measuring any side effect for current and future generations and forgetting most of the nonfinancial risks involved and the terminal ones at all. In fact, climate change presents significant financial and nonfinancial risks (on top of the existential ones) that are very difficult to gauge and estimate, let alone manage in a proactive or reactive way. Material reduction in emissions profile are required to deliver the Paris Agreement, and for all the talk regarding reaching net zero emissions, the overall equation between net reduced emissions of atmospheric carbon dioxide and its removal from the atmosphere is hard to balance and not without implementation costs and, more importantly, risks—even if done in a decisive but still too fast way.

Doing nothing is not even an option, from an existential standpoint. But also doing too much and too soon could be unbearable, with the risk of a rapid decline in asset values (equity and debt) linked to fossil fuel and producing multiple stranded assets (real estate included), as an unpalatable alternative to the physical destruction and economic disruption coming from extreme weather. From a logical standpoint, as already mentioned, two main risks have been identified by regulators and policy makers. On one side, physical risks could be defined as the potential negative events and bearings that are coming from the interplay of climate-related hazards. They could impact humans directly or the things they possess with some value or utility—the obvious example being a real estate property: buildings can have severe damages and infrastructure grids and services can become stranded or just get overwhelmed by excess demand or utterly damaged by the wild forces of nature. Even the intangible natural capital and many beauties of a place can get depleted (think of sky resort or rivers and lakes that are drought by heat waves, or even oceans that are tainted and spoiled by oil leakages) (Klier 2020). Physical risks can make the world not just un-investable and un-insurable, but also (and more importantly) un-livable. For example, extreme weather conditions can hamper the way we work and spend our leisure time—hence depressing real estate values and the overall attractiveness of a place. Food systems and global supply chains can also get wrecked and dislocated, causing a big city to wreak havoc, as rural areas are getting unable to feed its millions of citizens and causing sudden migrations, leading to further depredation and all-out wars.

Potential implications are huge, as pointed out by Daniel Klier. Between 0.7 and 1.2 Bn people are currently living in areas with a 14% average annual likelihood of lethal heat waves (and what about spikes in the fat tail of the distribution of temperature?). Also, compared to recent years, there is an expectation of 2X–4X the amount of capital invested in real estate and infrastructure as being at risk for damages from riverine flooding by 2030 and 2050, respectively. Finally, 45% of earth's land area is projected to experience biome shifts, impacting ecosystem, livelihood, and habitat, with huge people migrations to come, impacting the value (if not the existence at all) of the main metropolis impacted. No matter how physical risks are going to impact human lives, they also need to be analyzed with a financial valuation approach to get to the economic cost (expected and unexpected) associated with those events (e.g., storms, floods, or heat waves that are increasing their frequency and severity because of changes in climate—changing in their distributions' averages and volatilities). Proper, required investments need also to be assessed—for example, for the structural changes in the current stock of real estate that can make this more resistant to the new, expected spikes in bad weather or for the development of a reinsurance market that can help support the transition to a greener and more sustainable path to development and economic value migration.

On the other side, transition risk could be defined by the potential negative outcomes that are driven by the uncertain financial impacts that can result from a rapid low-carbon transition, for companies, sectors, cities, or entire countries. These negative effects could be financial (e.g., as an energy utility company shifts too fast from its oil-based offering to a renewable-energy one), but also consider reputational impacts (if done too slow and hence harming the equity brand of the company and the customer perception of it), technological breakthrough or constraints (if new technologies become available making the current obsolete or vice versa), and the whims of the end customers—we may all shift to vegetarians, in reply to climate change enhance sensitivity, or stop buying plastic bottles, making livestock and plastic bottle inventories potential stranded assets. These risks are also closely interconnected, with an obvious trade-off at play: the stronger and more immediate

Table 4.2 Humankind behavior can be seen as a key determinant of climate change risks (both positively and negatively)

Human behavior and climate change risk interception - Conceptual framework

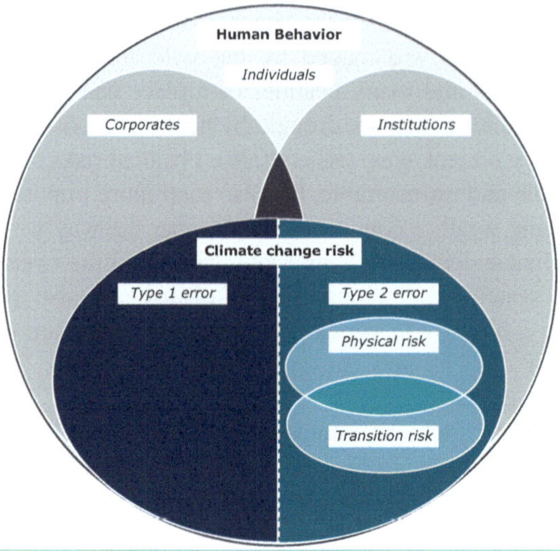

the action is taken to mitigate climate change, the greater the transition risks (with reduced physical risk as a counterweight). The more delayed and weaker the action is taken instead, the higher and potentially catastrophic the physical risks. Eventually, delayed actions followed by strong ones in an attempt to catch up would probably lead to higher physical and transition risks. These two main risks can be found everywhere in cities and embedded in everything we do, when working or spending free time—as much of the final action (investment, trading, or consumption) is happening in urban areas anyway and as everything related to climate change is eventually anthropologically based—driven by the behavior of individuals, corporates, and institutions as conceptually shown in Table 4.2. We mostly eat beef in cities and drink from plastic bags over there, no matter where the cattle is grown or the plastic bottles are produced. Hence the relevance of cities becoming sustainable to make sustainable the lives of the vast majority of humankind.

4.8 Green Is the Color of Money

Green is the color of money, an old saying is suggesting, paying tribute to the prevalence of the US dollar as the main currency of reference in global finance. Alas, green is now becoming the color of finance and not just because of the US buck.

Green finance is in fact becoming a bit of a buzzwords in banking, with regulators and policy makers, shareholders, and bondholders, not to mention end customers, asking lenders and asset managers to "get greener or die." Central banks, that were used to ask bankers to predict their capital shortages on the basis of the expected GDP growth, unemployment rate, or inflation, are now tasked with gauging the impact of scenarios—on cities, economies, and societies as well, as driven by changes in temperatures—as the Bank of England is doing in 2020 and with the European Central Bank following suit. Based on different carbon emission paths, raises in global average temperature could go from well below the Paris targets to a burning hot 4–5-degree ones—for each of them we may expect huge implications on either physical risk and transition risk (or both) for global cities and local ones, and rural areas as well, because of the trade-off between the two. The more rapid we shift to green investments, the greater and quicker the build-up of stranded assets and failed companies will be—it will be a very bumpy road as it won't be easy to switch off our addition to plastics or reconvert oil and gas majors to wind and solar renewables only. Would we commit to do without beef and cheese for life? Or avoid long haul flights, if not powered by electricity or making up with a weather balloon? The more lenders and asset managers will wait before taking action, the higher the physical damage, with an extinction risk looming at the end of most catastrophic scenarios. But also, the more lenders and managers will accelerate with the green conversion, the greater the pain will be felt in the economy and on their balance sheet—with billions of clients and voters that will be unhappy and hungry on both ways.

It is not a small responsibility to be a financial institution in these days—not just because of the unprecedented economic recession due to COVID-19, but also (and maybe, more importantly) for the role they are being asked to play, and the task they need to perform to pursue the "Good" for the environment while keeping doing "well" for themselves and their customers. Lenders and investment professionals are in fact being mandated by regulators to ensure that their capital, multiplied into loans and indirect investments via their lending and asset management businesses, is redirected towards companies and initiatives that are contributing to change the current climate change path and make the world and its cities a sustainable place again—while, at the same time, ensuring their solvency, paying dividends to their shareholders and income taxes to the government, and helping their clients in managing and making money. If we take a holistic view, at world level and the way it is polluting itself out of sustainability, we have argued that this "re-direction battle" needs to be fought, first and foremost, at cities and metropolis level. The design, development, and regeneration (from refurbishment to restructuring, from repurposing to rewiring of critical physical mobility and digital connectivity infrastructure) of global metropolis and main regional and local cities is a key issue that needs to be addressed quickly and decisively. For this, certainly, finance and real estate (and technology), as we argued in another book (Scardovi 2020), have a great and synergistic role to play. Finance can act as catalyst of this green-sustainable transformation of cities, providing the right incentives and means to ensure the optimal allocation of scarce resources towards that end. This is why even central banks and regulatory supervisors, on top of policy and lawmakers, are re-evaluating and extending their role as guarantors of financial stability (as impacted by climate change risk and covering the three main pillars (see Table 4.3)) but also of the sustainability of the overall, environmental ecosystem—because of the role financial intermediaries can have in directing (or diverting) financial resources on less (or more) polluting activities and contributing to the most efficient and effective, but also sustainable, allocation of scarce resources in the economy.

This is certainly a big responsibility for banks and other financial intermediaries, but it is also a big opportunity to do well on top of doing Good, not just with regard to managing the environmental impact of CO_2 but also, more extensively, covering a broader set of sustainability targets, often summarized and referred by the ESG acronym—of which the "E" of environment (S stands for "social" and G for "governance") is maybe the more relevant (as described in the Table 4.4).

The quest for the "E" is then bringing with it a potential big opportunity left mainly for green finance to address. As an example, green bonds have been growing at CAGR of 74% in last 6 years with USD 270 Bn in 2019.[3] Over USD 100 Tn investment in big cities-related "smart" and sustainable infrastructure (including transport, water and sanitation, telecoms, power and electricity transmission and distribution, primary energy supply chain, energy demand, and efficiency) are also required globally to achieve an "under 2°" scenario, with 50% of it that needs to happen in Asia (and with 16% in Europe).[4] There are a number of public and private initiatives accelerating but still subscale. More recently, for example, HSBC, one of the largest global banks, has announced the creation of a new unit focused on ESG solutions for clients and with particular focus to climate change to "help clients around the world rebuild and transition their businesses and economies in a more sustainable way post COVID 19" and with an aim (according to Daniel Klier) to become the "leading bank for the transition to a net-zero economy, helping to build a future in which economic growth and sustainability are well aligned." This is a huge opportunity to develop sustainable finance (as defined by financial services that are aligned with the Sustainable Development Goals, including the implementation of the Paris Agreement on Climate Change), with investment commitments as sustainable and impact investing assets at 26% of AUM in the United States in 2019.[5] For all the talk regarding smart cities, cyber and meta and ultimately "sustainable," a corresponding support of sustainable finance needs to be in place to make this transformation come true. The issue is, however, as discussed, how to measure, analyze, predict, and then manage the cities' unsustainability-related risks, starting from climate change and their quite unpredictable nature.

---

[3]Green bond highlights 2018, CBI 2019.

[4]Investing in climate, investing in growth, OECD, IEA, July 2017.

[5]Report on US sustainable, responsible, and impact investing trends, US SIF, 2018.

Table 4.3 Central Banks are reevaluating their role as regulators and guarantors of financial stability and their role in climate change

Tasks	Implications
Assess the size of climate related risk in the financial system	• Develop new analytical tools and conduct "climate stress tests" on physical risk (need to reconcile projections over several decades with the relatively short-term period considered under traditional stress tests) and transition risk • If climate stress tests find that climate-related risks are material, systemic capital buffers could be applied to mitigate the exposure to climate-related risks
Ensure that climate-related risks are well incorporated into individual financial institutions' strategies and risk management procedures	• Financial institutions should better understand climate-related risks and consider them in their risk management procedures and investment decisions, as well as in their longer-term strategies with reference to Basel framework: ✓ *Pillar 1 on minimum capital requirements:* If being exposed to climate-related risks is seen as part of financial risks, then it might be appropriate to consider capital requirements to reflect such risks ✓ *Pillar 2 on the supervision of institutions' risk management:* Regulators could prescribe additional capital on a case by case basis, *for instance* if a financial institution does not adequately monitor and manage climate-related risks ✓ *Pillar 3 on disclosure requirements:* Supervisory authorities can contribute to improving the pricing of climate-related risks and to a more efficient allocation of capital by requiring more systematized disclosure of climate-related risks.

Source: The Green Swan – Central banking and financial stability in the age of climate changes

AlixPartners

Table 4.4 Climate change risk is one of the components of the "E" of ESG, and its relevance for banks can be subdivided in two dimensions

E nvironmental	• Recent push for Financial Institutions to steer their activities towards being more environmental-friendly and responsible has led to different way to "bank": 1) Banks can support the transition towards a greener economy by allocating their capital accordingly 2) Banks need to understand how climate change is going to impact their balance sheets through increase in the NPLs and stranded asset write-offs	
S ocial **G** overnance	1 **New way of assessing and pricing of opportunities** • Changes in climate policies, with a potential significant carbon tax and further regulation, can impact the way banks do business • Understanding the role of the natural recourses and technologies in the transitions and aligning portfolios to that vision will be advantageous medium-to-long term	2 **New forward looking way of assessing risk of existing and new portfolio** • Existing portfolio has been assessed according to traditional methodologies and need to be evaluated considering new forward looking risk metrics • Companies' PDs can be adjusted for the industry-specific impacts of climate change

AlixPartners

4.9 A Big Data Approach for Climate Change

Big data and the use of innovative applied analytics such as AI/ML could help in addressing the issues related to the continuous monitoring, valuation, and prediction (and then active management) of climate change-related risks—helping banks and financial intermediaries in their financial risk management challenge and cities in the design and planning of their urban development and regeneration programs at safer standard levels. They could also help in devising more effective and efficient ways to operate them, with lower or zero contribution to carbon emissions. The use of big data and AI has long been discussed for doing better predictions and forecasting. Still, their use for making a city more environmentally savvy may look like a bold statement if not considered at the light of the several and significant constraints that more traditional risk management (and hence capital management) approaches would need to face anyway. In fact, at the basis of all this, there is the understanding of how physical risks could have un-linear and potentially snowballing effects in the economy (not much dissimilarly to what happens in a global banking and credit risk, with systemic effects spilling over across all sectors and domino-like among lenders). Also, the development of new analytical tools and the need to conduct "climate stress tests" on physical risks would need to reconcile projections over several decades and with the relatively short-term period considered under traditional stress tests analysis and include multiple endogenous factors (assuming climate change is, as discussed, anthropologically based), on top of a myriad of exogenous ones.

For example, the variations in the temperature degrees caused by climate change could be used as anchor points and the challenge of limited data history and of the inherent nonlinearity of the ensuing catastrophic scenarios could be addressed via scenario analysis. But then, the endogenous impacts of banks and asset managers pursuing "greener" lending and investing and of policy makers imposing new rules and regulations would also need to be predicted and factored in the risk valuation equation. It would be close to impossible to derive then a potential impact on physical assets and the health of people. Not to mention transition risk, with its business related implications, that is potentially even more complex to map out because of its many interrelationships with people (which are customers and end users), companies (that are changing continuously their green transformation strategies and with regard to many other factors), and public institutions (including policy and lawmakers and regulatory bodies). A big data approach would instead start by gathering all kind of structured and unstructured available data, leveraging the extended IoT networks that are already becoming part of smart and cyber-meta cities and will be quintessential to make them sustainable, also from a climate change perspective. In fact, the aggregations of structured and unstructured information are useful to qualify, position, and segment national and international climate change-sensitive areas of exposure. For example, let's consider the climate change risk that is impacting on the value of the real estate assets that are used as collateral to mortgages and development SPV (special purpose vehicles) and that are obviously representing a sizeable portion of the wealth of the nations and of its citizens. A simple, but critical, formal information (its address and geographical coordinates), for example, would be enough to position each and every property in the world map. On this basis, the relevant weather-related extreme events could be cross-referenced and geo-localized and monitored consistently and continuously, given changes in weather conditions, and ideally form the basis for ex post remediation and ex ante contingency plans. They could be also marked to market, indexed and insured or hedge via derivatives.

That would regard both physical and transition risks as shown in Tables 4.5 and 4.6 on the physical and transition risks involved and the potential opportunities coming from a big data, AI-based analytical approach. A bank that is lending to real estate investors and developers working on high-risk areas (because of flooding or hurricanes or just because of heat waves and raise in droughts and humidity) is exposed not only to losses on the value of the collateral because of physical risk (the building being severely damaged or destroyed) but also to losses on the properties' commercial value and related businesses that may be driven by the reduced attractiveness of the place, as it becomes progressively "unlivable" by people or less supportive of tourists related activities (e.g., for a hotel in the Alps, whose pick season is the sky one in the winter, if snowfall is progressively reducing or vanishing in the area). Therefore, it would be critically important to monitor, assess, and consider which kind of weather-related risks could be impacting on which area and on which properties, taking into consideration (at a more granular level) the

Table 4.5 Climate change risks and related economic impacts assessment requires innovative data and statistical approaches

Type of risk	Description	Economic impact	Implications
Type 1 error risk	• Provision of funds (either as lender or as investor) to "eco-unfriendly" businesses and counterparts	• Review of lending/ investment policies • Real-time monitoring of borrower's behavior and application of extra charges/ discounts based on "green indexes"	• Historical, structured data are not enough to analyze climate change • Multiple, real-time, and unstructured data sources (big-data like) should be tapped instead and quickly tested for relevance, by randomly generating "millions of ideas" • Traditional statistical approaches would not be able to manage these "millions of ideas" (potential independent variables) in order to understand multiple and highly interconnected climate change "cause – effect" un-linear relationships
Type 2 error risk	• Lending/ investment exposures that could be jeopardized by negative weather events or by the transition towards a greener economy	• Identification of key climate change drivers in: – Changes in weather conditions – Likelihood of occurrence of extreme negative events	
Physical risk	• Loss on lending/ investment exposures resulting from the occurrence of extreme negative events	• Association of an expected economic loss/ lower economic performance to the: – asset itself (if the case) – bank loan portfolio (as lender) – investment portfolio (as asset manager)	
Transition risk	• Loss on lending/ investment exposures resulting from the transition towards a greener economy		

AlixPartners

Table 4.6 AI can be leveraged to assess and incorporate the impact of climate change risks into real estate assets' valuations

Suggested pilots	Potential approach	
• Real Estate is the largest investable asset class and, by value, much larger than all the financial ones (residential real estate: $150 Tn), not to mention infrastructure	**Mark to market valuation**	• A mark to market approach to Real Estate includes: – the analysis of comparable assets at single location level – the analysis of supply and demand – the analysis and monitoring of the attractiveness of the place (e.g. business and leisure)
• The attractiveness of a city (with urbanization planned to lead 75% of the global population to live in city dwellings) and the value created by its main companies and Institutions drive the available income and creditworthiness of households and individuals in turn impacting the valuation of Real Estate assets	**Physical risk**	• Physical risk is going to become a key determinant of NPL "cure rate" from collateralized properties with impacts on: – operational risks (e.g. leakages) – cost of maintenance and repair (e.g. landslides) – marketability (e.g. Grade A, liquidity of local market of reference etc.)
	Transition risk	• Transition risk is going to become a key determinant of global cities attractiveness, from a B-Leisure (or "wealth-being") point of view, driving huge value migrations across regions and places, hence impacting on Real Estate, hospitality and infrastructure business

Climate change as driving the value migration across cities and regions and as reflected in the changes in real estate value should provide very meaningful info-telligence on how the overall financial risks are migrating because of different geopolitical fortunes

AlixPartners

safety standards of each of them or how they are set up to cope with (for example) extreme temperature and humidity conditions. These set of information would then need to be correlated with each other, at a single asset and with a portfolio view, and stress tested, to take into consideration the concentration or diversification effects vis-à-vis extreme tail "green swan" scenario events.

4.10 Green to Death

Big data and the use of innovative applied analytics such as AI/ML could help in addressing the conundrum related to the measurement, analysis, and forecast of climate changes and of its related physical and transition risks on the city. Indeed, this is not a small idea and entails a great challenge in the overall pursuit of designing, developing, and transforming cities into smart, cyber, meta, and sustainable ones. In fact, to make it sustainable does not only mean making them greener and greener, at the risk of forgetting all the rest. For all the talk of sustainable and green finance, the decision of "going green" and invest just on initiatives that are reducing CO_2 and lend to counterparts that are ecologically supportive of the environment is a very simplistic and likely wrong approach. Certainly, lenders and investors may keep throwing money at counterparts that are killing the environment and producing CO_2 without considering its critical consequences—our "type 1 error." But they could also do just the opposite, on the basis of the green indexes and ratings now offered by multiple third parties, often on a very subjective if not scientific basis, hence just window dressing ("green washing") this "type 1 error." But even assuming they are doing this on the basis of more granular, objective, and scientific and updated data, simply throwing money to green initiatives and counterparts could just imply a wave of defaults among corporates, followed by disruptions in the supply value chains of many basic products and services (from mobility to food, from shelter to heath), likely followed by social unrest, the disruption of the fibers holding communities together, and the overall collapse of the global financial system, if not of entire democratic systems. It would mean killing our cities and (more importantly) our civilization for the sake of going all-out green.

Could we consider a world without oil from tomorrow? With no cars going around and cold winters for families to overcome (even burning wood is causing massive CO_2 emissions) and unbearable summers without air conditioning. Our ways of living would change dramatically and become very unproductive—most of our manufacturing and distributing activities could get to a halt, with severe shortages of basic goods (now made of plastic, a derivative of oil) and short stops to consumptions. International trade and commerce would also get to a hard break (as flights and boats are also using oil), with other sectors (e.g., semiconductors, used to make PCs and smartphones, TVs, and all other king of computing machines supporting our economic system) also becoming temporarily imbalanced or terminally impaired. Even the luxury and fashion industries would be impacted as they are still relying pretty much on oil. The same would apply if we change our way of using arable lands, agriculture, and farming to feed our cities—we could save soil from erosion but killing cattle and even people being fed by them in the process. In short, the pursuit of sustainable cities and hence sustainable countries and civilization—and the very survival of the human species—needs to rely on the way we account for all kind of drivers of unsustainability—including the pollution caused by CO_2 emission but also the too fast and too drastic disappearance of it, if we pursue a naïve transition to a "green to death" way of living. And it needs to rely on the way we measure and analyze all these causes and related effects, ideally in real time, to monitor, if not "now-casting," the current trends and emerging patterns. It needs, finally, to quantify and make as objective as possible these two risks and their interrelations, so that they can be indexed, measured, priced, and eventually insured, swapped, or hedged, with an important role to play for public institutions.

If even "traditional" physical risks are becoming impossible to estimate and even measure, they will also become uninsurable, hence damaging the economic and social environment of cities way above the actual physical damage done to properties and people. If transition risks, impacting all kind of sectors, counterparts, and even locations, are also impossible to measure, let alone predict, the cost of providing funding to companies will also become sky high and not necessarily leading to the optimal allocation of resources that is supposed to seek, via market dynamics and the mechanism of pricing. If both are not quantified and there is no way to make them comparable and with a good understanding of their inherent trade-off, then there will be no way of optimizing their combined effect, which will be left to chance, if not to chaos. In short, if "too green, too soon" may be the best recipe for disaster, "too sustainable finance, too soon" (if we think of sustainable finance as the one focused on avoiding and minimizing the type 1 green error and with a quite blundered approach to return analysis) could also cause disastrous effects, with polluting companies transitioning too fast to death and entire cities also greening too quickly to become habitable by wild animals only. The "type 2 error" should then also be taken into account and considered when funding is allocated and when projects leading to the transformation of entire sectors or locations are designed, selected, and implemented. They should both cover the physical and transition risks and their inherent, unavoidable trade-offs and should price them, to make them financially tradable, insurable, and hedgeable. Sustainable cities could then be pursued and realized with the help of "sustainable finance"—but with a different and more extensive definition of it—a sustainable finance which is considering and optimizing both type 1 and 2 errors, creating the basis for an efficient financial market, as a first step to manage and act on them proactively and effectively, considering all kind of digital and physical dimensions that could play a role to that end. Also, in this case, big data, IoT, and AI/ML are coming to help, to mitigate the overall climate change risk. To master extreme weather-related events, a cloud could be of help as well—albeit a cyber-physical one and not brining any rain.

4.11 A Carbon Tax or a Green Derivative?

Cities, as complex ecosystems comprising both individuals and companies, will have a great role to play to drive a sustainable "green transformation," optimizing the inherent trade-offs between "type 1 and 2" errors, but also between physical and transition risks. The most successful in this will not only become more sustainable, but also more attractive and hence able to drive wealth creation by competing at best in the global arena. But also, financial systems and capital markets specifically could give them a hand—and to large- and small mid-sized companies as well, in executing this "green transformation" for good while not getting out of business and die. In fact, assuming we could analyze, understand, and even predict in a fairly objective way the emissions of CO_2 and NO_2, there may be some ground to set up a transparent, real-time, market-based pricing based on a totally objective Green Index that could be associated to them. Talks on a global carbon tax and the killer idea to curb CO_2 emissions and even fund the transition to a green, sustainable economy have been in the cards for years, with coordination efforts still underway and becoming part of much larger negotiations (and related, tense retaliations) among the largest and most powerful geopolitical blocks—but not only, as poorer counterparts can also carry a significant weight on this topic, if we just consider the role of the Amazon region in the global ecological system. A global or a set of supra regional carbon taxes could support the right incentives for companies to transition and gather financial resources that could be deployed to fund the conversion or remediation of unsustainable businesses—it would all be done by the "public hand" of policy makers instead of the "invisible one" of the markets though, with all the potentially negative consequences of this. On top of this, other structured solutions could however be built and deployed. For example, Axel Weber, the Chairman of UBS, has famously stated that if we could build a derivatives business on the carbon tax, we could potentially solve the problem of green transition, with contracts like future and swaps that would help companies in managing the financial risks driven by this tax and by its overall impact on their transformation plan. Still, this would rely on the tax decision-making process and to notoriously fuzzy negotiating process leading to that (something that even AI can hardly predict).

A more holistic interpretation of Axel Weber's point could instead be if we structure a derivative business, linked to an objective index (e.g., CO_2/NO_2 emissions in Italy, or broken down by macro region), there will be investors that are "climate change skeptics" and willing to sit on one "short" leg of the derivative (this will be priced at zero on day one, with a 50–50% chances that will go either way). Investors who are "climate change savvy" will instead sit on the "long" leg and asymmetrically so. In fact, "skeptics" could just be insurance companies and lenders which have to pay higher P&C (property and casualty) and health claims or face higher credit expected losses if climate warming does happen. They will thus be "happy" to pay money to the counterpart of the derivative if climate change does not happen (or happens in a lower amount given the set benchmark and its relative contractual parameterization). This kind of macro hedge could also work for public institutions (e.g., the credit insurance and public finance institutions owned by governments) as, if climate change happens, they will need to pay for their health systems, for potential damages to infrastructures, for required remediation programs, and so on. On this basis, if it does not happen, they will be happy to pay corporates (as key drivers of the achieved reduction of CO_2 and NO_2) for such a nice achievement. On the other side, "savvies" could be all kind of corporates traded on the stock market that will get money if climate change is lower than targeted and will need to pay for the opposite scenario, hence putting up a strong incentive to transition fast and effectively. They could also finance the transition with the expected payments from the derivatives and avoid producing too much pollution and just about at the right time. This market-based derivative business would then work as a kind of carbon tax but administered via derivatives that are traded on an efficient, regulated market and with many-to-many real-time professional counterparts active on them. Obviously, investment managers and lenders could incentivize funded corporates to trade on the long leg of the derivative, and most ESG-savvy corporates could show they have skin in the game on top of their high-level (sometimes rather fluffy) statements on the willingness to become "net zero."

So far, listed companies can (voluntarily) disclose greenhouse emissions according to three different perimeters (i.e., scope 1, 2, and 3 and with increasing degrees of fuzziness on the valuation and reporting). In 2018, more than 5.000 publicly listed companies (approximately 90% of the world stock market) have disclosed information on their greenhouse emissions, even if with a certain degree of subjectivity and diverging methodologies. Referring to scope 1 (covering emissions resulting from both the company's operations and by the companies supplying energy—electricity mainly) emissions, each year circa 10 Bn tonnes of CO_2 are emitted by publicly traded companies. But when looking at scope 3 greenhouse emissions (considering the extended value chain that refers to the single company—from the extraction of raw materials through its suppliers and on to its end customers and users), 220 companies account for 84% of the total carbon footprint of the more that 5.000 that disclose the information—if these 220 could be made active traders of publicly listed derivatives, they would already address most of the addressable market (or "problem to solve"). Of these 220 "heavy emitters" companies, about 76% are investor-owned, and if weighting the investors' stakes in the companies by the companies' emissions, it would follow that 250 financial firms a la BlackRock, Fidelity, or Vanguard already "control" approximately 86% of their total emissions. If these limited number of companies were enticed to trade on climate change indexes—with the moral suasion of policy makers and with the investment decisions operated by global asset managers—a derivatives market could be developed in a very fast way and effectively.

4.12 Credibility Matters

The search for an index, to incorporate other, nonfinancial considerations linked to sustainability, has indeed being going on for a while, with new, comprehensive ESG indexes sometimes introduced and touted as the Holy Grail and, most often, then highly criticized as "not adding up" and "mixed-up," often prey to the vagaries of complexity and the conflict of interest of subjective analysis. Leaving aside the complexity of putting together a single view of objectives that relate to the environment (hence, mostly focused on climate change and green transition, but not only) and social (including inclusiveness, wealth distribution, access to basic services such as education, training, and the health system) and governance (related to the best principles and organizational models to manage companies—already debatable per se and with ample variations between the western and eastern cultures, for example), even just focusing on climate change could be a daunting task to find a common index that could summarize the topic could enough and, even more importantly, could be mad available in a very objective, transparent way and in an almost mark to market fashion. Several rating systems have, for example, been designed and are currently in use to assign a gradc to corporates with regard to their CO_2 or NO_2 emissions, but, as no single measurement approach is available to capture what a company really does, these systems are also left to a high degree of subjectivity, from the company itself (that has all the incentive to undershoot its own estimates) and with lots of blurring boundaries and inconsistencies at the economic system level (e.g., all measurements and contributions would not "add up" to the one recorded globally for the system, because of the different "scope 1 + 2 + 3" extended contributions and because of the incentive to lie). A "carbon tax" initiative would face, at name level, the same issue of objectivity and "truth" and, when ongoing, would also be subject to the vagaries of policy makers and global coordination, on top of the ever-present incentive for corporates to lie, to look greener than they really are. With no credibility, and the risk of manipulation (even bigger if derivatives are attached to a given Green Index), an idea of a carbon tax" derivative could look great but would fall short on execution.

Conversely, what is readily available, fully objective, verifiable, and certifiable by independent third parties is the level of CO_2 (or NO_2) emissions at country or regional/local level. A set of Green Indexes and related derivatives could be structured around these "serviceable/indisputable truths"—observed by NGO and aggregated by stock markets or other independent parties. Such indexes would then have credibility and could be used as anchor points to gauge as climate change, and its associated risks are developing—derivatives could then use these as a parameter, to drive payments to their counterparts, based on the index future performance. Objectivity is however not enough to support the development of a liquid, frequently traded market, if some kind of predictive model can be developed (think of the help the discovery of the Black Scholes formula has contributed to the tradability of options) and if, on top, other more extensive models can also be used to understand what drives CO_2 across the economy and society, as a whole. In fact, big data and AI could support just that, addressing the apparently unrecognizable complexity and fuzziness of human behaviors, changing weather conditions and their impacts on physical and transition risks. Even more importantly, AI could offer a viable approach to predict changes in CO_2 and hence in derivatives pricing, on the basis of publicly available data—leveraging all kind of open sources available to crunch trillions of data and produce millions of hypothesis to search for serviceable truths. A climate change derivative, designed and set to price at parity on day one would then fluctuate in value (with pay offs and pay outs symmetrically calculated for its counterparts), based on the changes of the Green Index (calculated as the opposite of the CO_2 one), with AI-based model that could be used to predict the index forward looking behavior and also understand what is driving that (our dietary requirements, e.g., or the use of electric cars, e.g.). These models could also be supplied as SaaS (Software as a Service) by the stock market promoting the index, and, of course, single counterparts to the trade could use their own, private data on top of the public one and invest in more sophisticated predictive analytics—to make money on the derivatives but also, as a side effect, contributing to the creation of a greater intelligence and better understanding on climate change risks—making some of their findings "open source" as well to the overall society.

Finally, a third element—the willingness and the incentive for counterparts to trade on the stock market—would be required to create liquidity and sustain the development of the market, with numerous and diverse counterparts looking to act on both the supply and demand side of the trade. On top of being incentivized by policy makers and institutional investors, corporates could be incentivized by lenders to take a short position on the Index (e.g., on one of the leg of the futures' contract that would get in the money if certain CO_2/NO_2 reduction targets are achieved). They would then be incentivized to cut their carbon emissions and across their scope 1, 2, and 3, regardless of the actual (and largely subjective) valuation that would be available for these (the Index would capture objectively the CO_2/NO_2 emissions at systemic level. It would stand as money commitment on top of fluffy announcements, and it would help them in financing their transition plan towards greener value chains of production and distribution. Certainly, there could be the incentive for someone to "free ride," counting on the others to do good while keeping doing bad themselves. Regulators, policy makers, investors, and lenders could monitor and prevent this, but eventually, the most effective way of control could come from the private sector itself. The peers of a company and even the end customers could have a watchful eye on the behavior of single companies, as much as students have, when taking exams in a room without professors controlling who is cheating, as they know they grade will be "normalized" and hence dependent also from the ones of their fellow students. In a way, the success of the initiative would mostly depend on the credibility of the Index and, in turn, on the right incentive of market participants to trade on this.

4.13 Climate Matters

Big data and the use of innovative applied analytics such as AI/ML can help in addressing the conundrum related to the risk management of the physical and transition risks driven by climate change. They should be an integral part of the assessment of a bank's lending portfolio, of an asset manager's investment portfolio and, more broadly, of policy makers when managing the redesign and redevelopment of large metropolitan cities and urban and residential areas, to make them not just more productive and attractive but also sustainable. With specific reference to real estate planning, designing, financing, and investing, each and every key decision-maker, when assessing the merits of a new development or urban regeneration project, should consider its geolocation and the quality and attractiveness of its surrounding environment and how this is going to change because of climate change risk. The quality of a given location or specific area are in fact significantly influenced by their level of pollution and how this is going to impact on the health and well-being of individuals. This can be measured through traditional KPIS such as CO_2 or NO_2 emissions and the level of hazardous dust in the air and be monitored in real time, if appropriate IoT devices are distributed across and connected over the web, sourcing all kind of existing data to make them available for mining and distilling serviceable intelligence. Their behavior in time can then be analyzed against a number of other factors (such as mobility, or the use of alternative energy, or different ways to design and regenerate entire neighborhoods and buildings), to better understand any cause-effect potential relationships. The same analysis can be done at climate change risk level, by extending the measurement, in real time, of any weather risk-related event and their potential effects on properties (monitoring, e.g., the damages caused by strong winds or downpours) and businesses (focusing, in this case, on the swings in business performance of entire sectors or business model taxonomies, as explained by their different locations and changes in extreme climate events and taking into consideration a number of commercial, financial, and operational explanatory factors). For example, the attractiveness of a resort in the Alps can be reduced during the winter season, but allow a more extended summer one, with reduced costs for heating but likely requiring new investments to comply with more stringent safety standards for the building and surrounding infrastructure. Entirely new value propositions could also be open up, for areas previously neglected by tourism (say, Greenland in the winter) or by manufacturing industries.

In order to do this, any kind of statistical analysis would become quite complex, if using more than a few independent variables to explain the dependent one (the cost of risk, in this case), and would need some heroic hypothesis ex ante, to capture the most critical cause-effect relationships. On top of this, if we assume that historical data are of little use, given the structural change in trends, and that the future developments of all these extreme weather events are likely nonlinear and closely intertwined, we could rapidly reach the conclusion that any kind of statistical approach is going to be of little use. The use of open source big data and artificial intelligence are instead addressing most of these shortcomings. They could rely less on historical data and start aggregating huge masses of data from all the open sources available over the web and from IoT devices for analysis, using AI/ML approaches that do not necessarily require the ex ante explanation of which independent variables should be short listed and used to get to a greater predictive power. They would therefore consider a quasi-infinite number of cause effect possibilities and dynamically select the best ones. They would keep self-learning and self-adapting as nonlinear climate changes unfold and would update their forecasts on potential impacts, almost in real time, as new data are fed into the system.

They would then contribute to spot trends and pattern that, if not for the long term, would at least be able to "now-cast" emerging developments, e.g., anticipating in the short run negative events and allowing some kind of preemptive measures. This information and intelligence could then also be used to measure and monitor the "location's green-will," to change the overall sustainability and quality of the environment and promote a greener, less polluted and safer ecological system. Specific KPI could be used to set targets for both the private (more healthy lifestyles and habits, more energy efficient businesses) and public spheres (incentives for green mobility, promotion of more green areas, etc.). They could also be used as benchmark to drive climate change-driven taxation and green-will incentives. And, as we discussed, to also promote financial hedging or insurance coverage (if there is information, there is always a price) on risks that can be made manageable, when mutualized, or bringing the system to collapse if unaddressed in presence of market failures. In short, beyond adjusting their lending and investment policies to support resources allocation towards less intensive CO_2 and NO_2 emitting setups, banks and financial services should anticipate and predict the value migrations within those and their main players that can impact on their asset quality and sustainable profitability. Real-time data capture, and its cross-reference with current banks' loan exposure and perspective lending portfolio, could support them in having a more proactive and forward-looking approach towards their climate change risk/return profile optimization. For example, all the real estate collaterals to the loan portfolio of a bank could be cross-referenced with their geographical coordinates, to then link them with extreme weather-related risks and their continuous monitoring—to also suggest, ex ante, which kind of capital expenditure or appropriate insurance coverage they would require, to make their value more stable and safer over time. All this could be done in turn more effectively over the cloud, with third party providers managing the data pooling from multiple sources and running sophisticated AI analytics to get the best intelligence out of at scale information. For this "data pooling" approach to work, multiple stakeholders should be convinced of the "public good" nature of the exercise, given that so much is at stake, for the sake of the sustainability of meta cities and of their citizens wealth-being.

4.14 Climate Conscience

A collection of climate change-related structured and unstructured information could be funneled, via dynamic integration with multiple sources and through different techniques, into a "climate change data center." This is just another example of how data could be monetized and put to a serviceable use for the whole community. It is also an indication of how much is at stake for cities because of climate change and indeed is a critical goal to pursue—likely the most critical—to ensure their sustainability in the long run. Finally, it is also a strong example of how the convergence of the cyber and physical dimensions is critical to ensure effective answers to most urgent and relevant issues—be them of a societal and ecological dimension or of a merely economical one. This "climate change data center" would act as an information aggregator to derive and feed the AI/ ML model used to monitor climate change impacts on physical and transitional risk, starting from graphic views, KPI, and trend analysis to move (in the mid-term) towards more predictive approaches. Once a common foundation is accomplished, different patterns could be tested with real data and moved forward to implementation. Real data, continuously refreshed, would be used to develop the model. The cloud (or multi-cloud) solution would then act as a cluster of high-performance databases, all sharing a common data access layer that gathers and stores all the data, performs aggregations and data analysis, and could also serve multiple stakeholders via APIs. In a first phase, AI/ML could be adopted to normalize data and find correlation patterns among the variables (dependent/ independent), to then define the algorithms that will best relate the dependent variables (i.e., sea level) with the new innovative ones (e.g., IoT carbon emission)—the independent variables. Since many of this data is constantly changing, the model is always considering new and updated data and self-learning. The integration with climate change smart data sets and the following data analysis would then aim at finding correlations between climate change variables and then define a climate change risk-adjusted VaR (value at risk) that could be used to judge, plan, and compare.

The correlation found by the model would also aim to offer a more holistic, comprehensive, and forward-looking view over the "type 1" and "type 2" risk errors, helping the different stakeholders with a "parallel speedometer" that could be used to take decisions on the design and development of sustainable cities and to anticipate potential issues. In this way, also nonlinear, "black swan" ("green swan") events could be captured to also gauge the potentially seismic impacts of climate change on (for example) the value of a given territory, the real estate and infrastructure assets or commercial businesses of a given city, and the potential impact on the wealth-being of their citizens.

Whatever the "end-grade" scenario foreseen for a world that will become hotter and hotter, massive value migrations are expected across countries and locations, industries and business players, and entire societies. These migrations are almost impossible to measure and predict via traditional data and analysis, and big data and AI/ML analytics could give a hand. More importantly, these migrations will happen at the level of big, meta cities, with their climate change-related sustainability becoming a critical driver of competitiveness and attractiveness. It follows that, in our view, the ability to measure, understand, and predict this new kind of climate-related risks will all become critical capabilities, for most competitive smart to cyber, meta, and then sustainable cities. New ways of assessing and pricing the risk/return of lending and investing opportunities linked to the climate change transformation ahead are then required. They could be used to put a price on almost everything, to drive public policies (e.g., carbon taxes and hard regulations imposed on most polluting businesses), but also private ones, at company and individual level. What about, in fact, leveraging data and information, across the cyber and physical spaces to drive change in our private behavior, e.g., from the way we eat and get out of meat-heavy diets, to the way we boycott high polluting consumer products, changing our preference in mobility and transportation? Eventually, industry-wide and company-specific value migrations will be driven by changes in demand, in a free market and free society where we assume, we can ultimately vote with our feet and decide, in conscience and through our investment and consumption choices, the kind of city we want to live in.

4.15 An Anthropological Challenge

In fact, climate change is not happening in a vacuum and it's not exogenous in nature—it's nothing like the proverbial dinosaurs' comet hitting the earth ecosystem and coming from very far away. Nonetheless, it is a potential terminal event that requires the harnessing of the best info-telligence available for us to address this and survive—in a way it's a formidable challenge where our collective cognitive ability may just not be good enough and a little help from AI/ML could come as very handy (with Elon Musk Neuralink direct plug in or not). In fact, climate change is induced by humankind, hence endogenous and because of this potentially controllable, if enough resources are deployed on the basis of a thorough understanding of its main causalities. Climate change, as we discussed, whether related to our behavior as individuals, private companies, or public institutions and in an organized or mostly un-organized way, it all starts from the way we invest, produce, consume, save, and recycle—and mostly in the urban setting of the city (cities are generating most of CO_2 anyway, as producers and end users of correlated products—e.g., the beef that is mostly eaten in cities is produced by breeding in the countryside, with significant production of CO_2). According to this, climate change is first and foremost an anthropological challenge that is ultimately determined by our behavior, as individuals and as part of complex organizations such as companies and institutions and with both as part of the cities' ecosystem. It follows that, in a first step, the universe of big data to consider and analyze for the potential identification of cause-effects relationships is our own behavior—including in this our individual lifestyle (e.g., how much and the way we travel, eat, dress), companies' and institutions strategies (which kind of products and services they design and how they produce and distribute them), and public policies (including in this, on top of economic policy-making, also the overall context of the rule of law we use to manage our civilization).

If a first step is all about to measure and monitor, in a second step, the humankind behavior being analyzed could then be associated with changes in the key drivers of climate change (CO_2, but not only) and in a cross-section way, given limited data history availability and the opportunity to run parallel analysis in real time and across different human living ecosystems, e.g., cities—to identify what can make better and more sustainable a city in the climate change race. Key drivers of climate change could then be associated with short- to long-term changes in weather conditions and extreme negative events and, in turn, put into correlation with "physical risks" and "transition risk" and with an implicit trade-off between the two. In fact, physical risk can negatively impact on almost any kind of real assets (e.g., damage to properties) but also individuals (e.g., accidents, health-related issues), while transition risk can impact negatively on the overall performance (including regulatory fines) of companies, potentially putting them out of business, if they are not good and fast enough in readjusting their competitive strategy at the light of the upcoming changes in supply and demand. Finally, these risks (leaving aside the intangible and sometimes spiritual impact) could be associated with an expected economic loss, to the asset's owner, individual, company, etc. and then, indirectly, to the bank loan portfolio (as lender) or investment manager (as asset manager). The overall idea is to use big data and artificial intelligence (BD/AI) to describe these phenomena, start analyzing them to look for cause-effect relationships, and build visual dashboards and KPI and some predictive capabilities to evaluate, anticipate, and optimally manage these risks. Ultimately, these info-telligence could then be used to transform our cities into better one, almost by design, as not only all kind of potential risks are reflected in the development and management policies of real estate and infrastructure, but also the very same city is set up to optimize the anthropological behavior of its citizen so as to minimize the human-induced contribution to further pollution. As climate change is driven by the way we live, cities can help us in living in a greener way and live better in return. And big data and AI can help cities with the identification of serviceable truths and workable solutions that could be applied to and influence our anthropological behaviors and ultimate ontological quest.

References

Adam Green, Deep data helps cities prepare for disaster, Financial Times, Juanary 29th, 2020

Daniel Klier, The transition to a low carbon economy – risk and opportunity, HSBC, February 2020

Claudio Scardovi, Finance, Real Estate and Wealth-Being, Bocconi University Press, 2020

Chapter 5
Our "Other Normal"

5.1 The Day After

The so-called "new coronavirus—COVID-19" health and humanitarian crisis has been costing many lives and producing an almost unprecedented, in recent, peacetime history, debacle of global GDP (gross domestic product) and subsequent wealth destruction. Its unfolding, almost uncontrolled development and related impacts, in social and economic terms, have been dubbed as of almost biblical proportion—according to noteworthy commentators, including the past Governor of the European Central Bank Mario Draghi. In fact, it has been damaging all major geopolitical areas—developed, developing, and underdeveloped alike. It has been hurting the lives of billions of households and the fortunes of millions of corporates and SME (small medium enterprises). It has brought death and misery—directly (through the illness itself) and indirectly, forcing people to prolonged "lockdowns" and companies to extended "shutdowns" that has put many out of business. In a way, it has shown, in all its subtleties and powerfulness, the extreme frailty of very complex, and otherwise super-efficient, global supply-and-demand value chains that are currently linking rural areas and cosmopolitan cities across countries and extended regions—all working with thin buffers and limited redundancies and with just in time production-to-distribution life cycles. It has shown the weaknesses of interconnected systems working in real time, even if digital-wise and innovation-prone. Finally, it has shown the frivolousness of many of the things we deemed (falsely) essential, and vice versa it has reminded us how many things that we are giving for granted are, in fact, quite essential. It is not an exaggeration to say that a single pandemic has been able, if not to turn upside down our inner beliefs, at least to make us rethink them in a very thorough way.

Whatever the evolution and return-to-normality of this "new coronavirus—COVID-19," we may already abstract some lessons learned and ponder on what this really mean, for the future of large sways of humanity living in urban areas and in metropolitan cities that are aspiring to be "smart." As a first, we could argue, a

© The Author(s), under exclusive license to Springer Nature Switzerland AG 2021

C. Scardovi, *Sustainable Cities*, https://doi.org/10.1007/978-3-030-68438-9_5

number of lessons learned have been related to the monitoring, early spotting, and short-term extended controlling of the virus, for this and the unavoidable new ones, to ideally avoid its evolution into epidemic and then pandemic stages. As a second, a number of lessons should be drawn on the best way to ensure an orderly and well-regulated and controlled "lockdown and shutdown" process, in the context of large areas and, more importantly, in cities with several millions of people and buzzing of social and economic activities—where democracy and freedom are paradoxically making this task harder and posing relevant ethical questions on a number of trade-offs (how more health is more important than less freedom? How health is driven by the control of the pandemic and the avoidance of a general starvation and impoverishment of the society?). As the world is getting ready to finally turn towards the "day after" of this pandemic, a third lesson could also be learned on how to manage the "un-locking/un-shutting down" (or "re-opening") phase, with its long-term aim of returning people and companies back to normality, quickly but also safely. Finally, and maybe more importantly for the sake of this "sustainable city" discussion, a fourth lesson could be inferred (as it will take time for this to fully unfold and develop) on the long-term changes to our ways of living and working—in a new, other normal that will likely be very different to our pre-COVID-19 one and that could possibly (this could be the piece of positive news) be better suited to manage the next calamities to come—climate change, first and foremost, could be something we will need to face and address soon, a much greater challenge vis-à-vis COVID-19.

Our hypothesis, for discussion, is that the events related to the COVID-19 health crisis have shown even more how the digital and physical dimensions could and should be linked together, to become fully intertwined and enmeshed in an hyperconnected environment and work together as "cy-phy," to improve our safety and resilience, efficiency, and efficacy—hence ensuring the sustainability of societies and economies and of big cities as their best and most obvious proxies. The COVID-19 crisis could in fact accelerate the adoption of digital technologies that are continuously capturing data from real life, cross-referencing them with "twin worlds"—pursuing health, security, economic stability, and so on. It has also shown how "smart" services cannot be limited to basic utilities but should be always referenced back to households and companies. It has also shown how "big digital" can be critically relevant, but also worrisome and potentially uncontrollable because of their grip on info-telligence—as this, we have been reminded by the pandemic, is also, at least partially, a not renounceable public good. And with regard to cities and real estate, it would be too simplistic to say that everything will change (to eventually remain the same). Home with gardens are reporting a pick in demand in the United Kingdom, as reported by the FT in September 2020, with suburban areas also becoming more fashionable and the crowded offices in the city and Canary Wharf becoming less and less so (and still pretty empty, no matter what the incentives of the companies and government are). All these is certainly relevant, but the "physical to physical" swap is just the beginning—the change that could be pursued not to change at all, a là Prince of Salina's way. In fact, the future is cyber-physical.

5.2 The Death (and Rebirth) of the Office

We already discussed, in Chapter 3, some of the potential, first-order implications of the pandemics on real estate—in a simplistic way. The cities may not be dead yet but maybe offices are[1]—have written well-reputed authors. Several companies, from Twitter and Facebook but also more traditional ones like PSA, are offering to their employees to keep working from home forever—it is good for the pandemic, but also for climate change and potentially more productive, as time travel is cut to nothing and the division and collaboration of intellectual work can mostly happen online. It could also be good for our health and families, for raising children and living a less stressful life. Melvin Webber, a futurologist, had already predicted (even before COVID-19) a post-cities era—with offices being crowded out by chalets in the Alps or by lakeside or seaside villas. Still, there may be other positive factors that we could miss from the buzz of the city, apart from its frantic traffic, noisy overcrowding, and insane cost of living. Indeed, according to Mark Granovetter, a sociologist, all kind of societies are based on "strong relationships" (close ones, e.g., among a group of well-interconnected friends) but also "weak ones" more based on chance (where we are in touch with a multitude of more diversified and heterogeneous people and almost on a mostly "by chance" basis). Apparently, innovation (and serendipity within this) is easier for the second kind of relationships. These cannot be planned or structured in advance with, say, a scheduled zoom between a poet, a data scientist, and an entrepreneur looking for a good idea to develop. They need physical interactions to happen, maybe at a bar, as they are mostly spontaneous and driven by chance—because of their physical nature. Maybe, in time, some of this social interaction-based "serendipity effect" could also be recreated on the digital realm (as websites are already managing sez, romance, and even marriages) but it will take time—no cafeteria effect is still fully available online.

The potential death and rebirth of the office are, of course, carrying a number of implications that are larger than real estate, or of the city itself, and with impacts on the future of work (and hence of our way of living). In fact, work in the office, once associated with routine and conformity, is now fast becoming a source of economic and social uncertainty—with macro and microeconomic implications and a tense political debate between the governments, the corporates, and the landlords. The pandemics, write *The Economist*,[2] is just revealing how many offices were run as relics of the bygone past—as many jobs were going to be impacted anyway by robotization and AI. As a result, COVID-19 may have just prompted and accelerated a long overdue phase of technological and social experimentation—neither business as usual nor a fatal blow to the office. A new "compact" may be needed to regulate "smart working" from home. But an even more holistic approach, involving employment laws, culture and behavior, and physical and digital setup, is required to ensure

[1]Carlo Ratti, Project Syndicate, www.project-syndicate.org

[2]Office politics, *The Economist*, September 9, 2020.

a new, upcoming, and targeted "sustainable working"—this will also require reimagining cities' centers and suburban areas. It may be argued that this could take tens of years and a huge amount of "war of attrition" between special interests, the natural inertia and unwillingness to change, and even the technical difficulties to redraw, almost from ground zero, entire neighborhoods and related mobility, multiutilities related, and digital infrastructure. In fact, PC have become widespread at the beginning of the 1990s, with cheap bandwidth already available 10 years later and Zoom and Slack already available another 10 years later (e.g., 10 years ago).

The pandemic, for all its bad, has unleashed all these potential drivers of change bringing work, the office, and cities center on the verge of disruption. A disruption is bringing pains but also huge opportunities, if the great amount of capital that is being redirected towards the reconstruction of the main nations' economies will be invested to design and build the sustainable cities of the future. There may be huge write-offs on "old" property assets and residential areas and (or) big opportunities for value accretion on redesigned and fit for purpose real estate and cities—particularly if governments, already short on capital to invest and on new, expansionary monetary tools, will instead focus on the streamlining of the red tape and the elimination of regulatory constraints, which are making today impossible the valorization of huge, country-level opportunities. Think of Italy and of its many natural and artistic beauties and the opportunities offered if, as in the incipit of another book of ours, Palermo could be turned into a new, EMEA-based Singapore, but without its humidity and with plenty of land and historical heritage to develop and capitalize? And if Milan could be turned into another London, as financial services hub and with a taste for the luxury sector and for the good living associated with the "Made in Italy"? Or if Rome, as political and artistic center of Italy, could be brought back to the level of grandeur associated with Paris?

The pandemic has suddenly accelerated to advent of a new, smarter future. However, as William Gibson, the science fiction writer, could have noted, this smarter future is just not evenly distributed. For the office and the white-collar workers, most critical technologies (Zoom, Skype, and Teams) have been around for years, but it was the pandemics and the subsequent lockdowns that made them essential and universally adopted. They (and their employers) are now worrying about the loss of community, given the limited amount of time they will end up sharing a physical space, even after the pandemic is gone and certainly, as we have argued, offices will need to be rethought and reinvented but will remain a fixture of future life. However, it is far less clear what the future for other blue-collar workplaces will be, given that workers spend most of their time in shops, restaurants, and factories (Susskind 2020). Indeed, the need to social distance people that are inherently more at risk for the pandemics (and, apparently, with an indirect relationship between risk and salary) will likely imply that, to protect their health, they will find themselves redundant and without a job before the arrival of a globally available coronavirus vaccine. Robotics has been, for example, one of the best performing sectors in the last few months, given the accelerated need and reduced breakeven point for companies to invest on automation and robotization, at the cost of fewer jobs. With more e-commerce and mixed cyber-physical sales (e.g., with physical showroom and sales that can only be finalized online), and with greater round the clock robotization of industrial production sites, blue-collar jobs will become more and more unsustainable, with social costs that will also impact the city and the need to find other ways to promote inclusiveness and cohesion. For all the talk on the "death of the office," all kind of other workplaces will likely be hit harder in the mid-term, if not on their real estate value (as buildings will be needed anyway to host robots) on their social ones—with many jobs that will be gone for ruined as well as for changed businesses.

All these challenges could start from the offices and the way the service businesses are working and accelerate the redesign and redevelopment of big cities, like Milan and Rome. And if the companies will adapt to a pattern of selected attendance in the office as meeting center and social hub, other suburban areas could also prosper, provided the right set of mobility and communication infrastructure—with complete overhauls required of currently in use complex urban-planning rules and allowing properties and urban areas to be redeveloped for new uses and ways of life, and likely for the better. As COVID-19 has forced a radical shift in working habits, many people (and the environment) have started to take a liking at them, and it will be increasingly difficult to bring back the beast into the cage, also for lack of arguments (most of us would now agree that "going back to work" does not imply necessarily "going back to the office."[3] And, talking of climate change as the next big challenge for humanity, cutting the level of traffic in the city is certainly something that can help to reduce pollution and help us in regaining some lifestyle rebalancing and well-being). However, a number of trade-offs remain and complex issues to crack as well, requiring untested approaches and, likely, the innovative use of big data, artificial intelligence, and much else.

5.3 Social (Distance) in the City

Eventually, a "cy-phy" approach could be a good solution to reconcile all this and allow for the rebirth of the office and much else in the sustainable city of the future. It will involve the redesign of old and new buildings and entire new concepts to work there (the "15 minutes" city being just one of them). They could also foster new ways of living socially and with a complete new set of values and behaviors—this may seem far-fetched and long term but in fact it is not, thanks to COVID-19. As we all know by now, the social distancing enforced through extended lockdowns across all main global cities has suddenly emerged as something totally unpredictable and unique. Also, their gradual restart of normal activities has also been looking like something unprecedented and untested. In fact, many things have changed in the way we behave and will be changed forever, as we will face new pandemic and climate change-related threats. The many answers to the pandemic, social distancing,

[3]What a way to make a living, *The Economist*, September 9, 2020.

more communication over the web, and less travelling, but also the quest for our return to normality (with more offline data capture used to build twin, virtual, network-based version of extended families and companies) look very much as precursors and accelerators of advent of the "cy-phy" city discussed in this book. The development of the interconnections of our digital and physical dimensions has allowed us to be present when we are not, and to support the safety of our health when we are, by checking our basic functional parameters and interactions with other people at high risk of being or becoming infected. Through the analysis of both digital and physical dimensions, a number of monitoring and reporting apps have become available and, because of the COVID-19 crisis, increasingly accepted and acceptable, at the light of the public good they seek to pursue—not to mention the digital and physical "first aid" support to our, and others', health that could be made available in case of need and thanks to digital innovation.

In this post COVID-19 new world, where the extended connectivity and the fusion of the digital and physical dimensions are playing a critical role, the "big digitals" have been thriving even more. Not counterintuitively, they are emerging, among so many big losers, as one of the few big winners of the pandemic crisis.[4] Not only large platforms such as Google, Amazon, Microsoft, and Facebook, with their e-commerce, cloud-based services, and digital socialite propositions, have seen their positioning become even stronger: Amazon, for example, which has kept hiring through the crisis as almost "forced" to take even more share from locked-down physical commerce shops; Google, which has contributed to support surveillance services during and after the lockdown; and Facebook that has seen messaging increased by over 50% in those countries that were hit harder by the virus. But also, smaller platforms, like Slack and Zoom, which help businesses to operate remotely, have been able to become well-known household names and the darlings of investors in the midst of a market crisis. They may all be living a great moment and positioning to become a key contributor for the design, development, and management of the "cy-phy," sustainable cities of tomorrow. They could also become their main detractor, if their weight and role is deemed too much to bear and too expanded to tolerate in a democratic, free society. Not surprisingly, new calls have been made for their breakup and regulation (or even nationalization of some of their services that could be deemed as "public goods" during crisis time), but for the moment surveillance capitalism (or a la "Chinese way") seems here to stay as a necessary evil (better a big brother than a deadly and uncontrolled pandemic). Sensitivity with regard to intrusion on personal data has diminished—understandably so, during a period where normal citizens have not even been allowed to leave their homes and take a walk. But it will get back, as soon as things will get back to (almost) normal.

Anyway, the role of "big digital" as info-telligence utilities has never been so directly exposed and obvious. Our social distancing disruption would have been even more intolerable without the e-commerce capabilities of Amazon, or the search

[4]Don't waste a good crisis, *The Economist*, March 3, 2020.

and information sharing ones of Google, or the socialite apps of Facebook. Apart from their services, they are also providing a formidable financial war chest—with plenty of capital and liquidity to splash in case of need. In times of almost unprecedented financial collapse, the main "big digital" (Alphabet, Amazon, Apple, Facebook, and Microsoft) have a total of USD 570 Bn of gross cash available—with their shares outperforming the market consistently since the beginning of the crisis. Vice versa, many small and worst funded fintech are suffering and will be ripe for consolidation (big digital could play as savior of last resort). As argued by *The Economist*, this may become an historic opportunity for big digital to propose a new deal for the citizens and with the policy makers and regulators, with clear and verifiable rules on how they capture data, use them, publish information, and leverage intelligence—ensuring the fair use of the content they publish and the fair treatment of the competitors that use their platform. It could also be an opportunity for them to contribute in a positive and open way to the design of the blueprint and development of the framework of the sustainable cities to come and of the "cy-phy" competitive dimension that will triage the winners and the losers of the future. On one side, they may face a scenario of post crisis breakup, "utilization" (with caps on profits and standards on offered serviced), and even nationalization. On the other side, they can foster a scenario where they become the kingmaker of the sustainable cities of the future, which are much more than "smart," and support the pursuit of the wealth-well-being of entire communities, across countries and regions and helping them in facing all kind of personal and professional routines, but also the most extreme challenges, including the existential threats to humanity to come. This is no small challenge and not something they could address without an ethical vision of the world to be. In fact, it is a challenge to support the move from old to "smart" and then "sustainable" cities.

5.4 Locking Down

The use of digital technology, in terms of big data, extended connectivity, and AI, has become increasingly relevant when countries have faced (and will be facing, in future waves) the need to ensure that compulsory quarantines is respected by people. Hong Kong (Hui 2020) and Singapore, among others, have been at the forefront of this, leveraging their extended connectivity and access to offline, public, and private data and innovative approaches to the use of technology (and—we could add, the greater acceptance by people of the State's "intrusiveness" into our everyday lives). The government of Hong Kong has, for example, rolled out electronic tracker wristbands that alert authorities on rogue behaviors—such as the avoidance of a mandatory 14-day quarantine for all the arrivals to the city from overseas. What initially was interpreted as an extreme measure for extreme, but almost unrepeatable, times has progressively become the norm as, for example, Singapore has started and rolled off, and then restarted, three different lockdowns in the space of a few months, given the resurgence of the pandemic and its target to keep it at bay. Digital surveillance can in fact support a draconian, initial effort, to single out the virus completely. Or it can be used to keep the total number of infected people at a stable level ("flattening the curve of the virus transmission"—to avoid the collapse of the health system and in the understanding that its complete suppression can only come through either a vaccine or through an "herd immunity."

From a technological point of view, a wristband works better than a smartphone, and everybody can (or be forced to) use it, babies included. It is, as a smartphone, unique and geo-localized in real time and associated with a unique QR code—then associated to an identity card. In Hong Kong, a wristband is given as a mandatory monitoring tool to all new arrivals at the airport. Once home, these people walk around the apartment to calibrate the wristband that will capture data and feed the algorithms of a monitoring app based on geofencing technology—something much more precise than GPS location tracking. Each home has then a unique set of communications signals, including a Wi-Fi network, Bluetooth technology, and cellular networks. As the user walks around the apartment, the app creates a composite signature of the home. If someone tries to breach the quarantine by leaving their home, the app triggers a warning and alerts the government. This technology—it is argued by the Hong Kong government—poses no privacy concerns because it merely looks for information to deduce whether someone is inside or outside of their home. But other wristband and wearable applications could do much more. By capturing information on the state of health of the person and its whereabouts in the city—not to mention keeping track of his activities and an audit trail of the other people (potentially infected) that he has met. Taiwan, for example, is monitoring cellular signals on phones, of people moving across the city. And Israel uses GPS data tracking to check on quarantined people (but with less precision and gap in coverage).

Singapore, likely the most advanced in digital connectivity at city-State level, is then employing a mix of these technologies and potentially offering a very extended coverage to trace confirmed cases and ensure their lockdown. Given the required level of information, this lockdown can be switched on and off at a tap of a finger, on the basis of their risk or of their function. Different levels of lockdown could then be enforced dynamically to diverse people and given the overall development of the pandemics. A risk matrix (measuring the chances of death for specific cohorts, if they contract the virus) could be set on the basis of age, gender, and other critical pathologies. Also, the probability of contracting the virus and, in turn, infecting other people could be related to the working sector, the back or front office activity performed and the level of mobility required ex ante or observed ex post on the basis of the tacking. This analysis, almost static, could then be matched by another one, considering the current health condition of the people, as measured—for example—by his body temperature, heartbeats, and sweat. A wristband could send these required information to a "twin city" pandemic monitoring center, to differentiate the degrees of freedom allowed to different counterpart and change them dynamically, should basic functional parameter change. This could be enforced in a dictatorial country, or simply (and temporarily) because of a regime of "force majeure." Even in GDPR compliant, democratic, freedom-sensitive countries, such an approach could be successful if this monitoring was based on the voluntary sign-off of the individual. During crisis times, where people are forced for month within the four walls of their home, it should be an easy trade to ask for a temporary sign-in, in exchange of the release of some of the severe restrictions otherwise imposed. And from this use case to other, broader, and extended ones, to cover other forms of potential illness of rehab behaviors, the steps would be small and merely incremental.

5.5 Unlocking

If the effort of locking-down billions of people may look paramount, the challenge of gradually, selectively unlocking those same people is, as we have learned the hard way, even more so. In fact, in an ideal world, this could be a "once and for all" task—certainly difficult, but with a clear path ahead. In the real world, however, it will likely require a dynamic approach and different shapes through time, as different waves of the same pandemic could reappear (with the target of keeping a "float" target of infected people at a stable, manageable level) or new calamities, all potentially impacting the sustainability of the city. As global cities are trying to emerge from the COVID-19 crisis and reach a new normal state, they will need to test-bed not only all kind of new technologies but also the very design of the new "cy-phy" city to come—not just smart and resilient and apt to adapt and survive through difficult times. The social distancing is bound to disappear neither suddenly nor completely and the return to work of important industrial sectors needs to be gradually phased in and differentiated by specific risk categories, setting the stage for new ways of doing things that will be here to stay after the crisis. Even schools and academies will not be the same for at least a few years. Mobility needs to remain partially limited for a while and selectively regulated with bottlenecks that needs to be absolutely avoided (for contagion) and resolved (for efficiency and efficacy reasons). Also, a rise in social unrest, thefts, acts of violence, and other forms of extreme protests that could arise because of lockdown and other social restrictions should also be considered and prevented and, where unavoidable, managed accordingly. Again, the ad hoc applications of such technological tools could then be extended to normal times and carry on.

At the same time, after a protracted social and consumption lethargy, a strong demand for certain services and products, either related to the backlog of activities accumulated in the past weeks or to the strong desire of the population to go back to the normal life, should also be planned and managed accordingly. We have all seen this pent-up demand producing queues and bottlenecks, made even more evident by the many global supply chains that will stand as significantly impaired when not broken. Think, for example, to the need for required bureaucratic procedures and certificates and all kind of non-necessary goods that have become out of stock at our homes and are then finally available. Or at the backlog COVID-19-related surgical interventions or other diagnostics impacting health systems still under some stress.

And at the request of services to the person (e.g., hairdresser) and goods (e.g., plumbers) that require physical interactions and need to be managed in a different ways and with different setups. After a forced lethargy, there is almost always a strong demand for off-site entertainment and training—from cinema and theaters, parks, and gyms. But any form of aggregation needs to be reintroduced very gradually and with close controls and following the implementation of new rules—again requiring a new city setup. A strong and urgent demand for companies and families to access any form of financial and non-financial support that may be available will also require support to deal with often unclear and not fully tested processes. Finally, on top of all these physical needs, psychological support will be needed to cope with the new normal—a further challenge for public and private institutions, households, and corporates as nothing of the past could be taken for granted and most of the future, our behavior included, will be different.

A "smart city" could provide some technological functions to help on that end—focusing on shared infrastructure and public services, to monitor traffic, for example, or pollution, safety, and so on. In Table 5.1, for example, it is shown how an IoT connected system could work to support body temperature scan and become "embedded" in the building, augmenting its safety. But it should be the focus on the individual, and the leverage and mixing of both online and offline data, that could be more critical in this restart of activities—more reconstruction and development in many instances than a quick "get back on track." The management of both digital and physical dimensions for all the needs and related activities described should be able to provide better support to overcome the unlocking challenge—acting dynamically and being able to feedback continuously info-intelligence, from atoms to bits and then back, to improve adaptability and resilience. In a way, the same "cy-phy" approach that can help cities in avoiding the next calamites and in managing down their effects when happening should also be the one that can help them in getting back to a new normal—leveraging data, extended connectivity, and artificial intelligence to produce serviceable info-telligence and putting this to use. Only in this way cities could become sustainable. Extraordinary events are in fact coming in waves and cities need to plan ahead for the next black swan and for the following new normal. Being smart in the current ecosystem is in fact of little value, if cities are not able to be sustainable in the upcoming ones—to survive and develop.

5.6 Restart in a Post COVID-19 World

The restart of the normal activities in a post COVID-19 world are certainly a case in point of the many challenges to come that cities across the globe will have to face. They also are, with all their failures and shortcomings, uncertainties, and stop-and-go approaches, a lively example of how cyber and physical could help us in making cities more resilient and sustainable, through good and bad times. For example, in a restart and then transition phase that could last anything from a few weeks or months to a few years, some basic operations need to be put in place, to ensure the resetting

Table 5.1 Use cases of thermal body scanner cameras for COVID-19 emergency and other applications

	Applications	Rationale
Thermal body scanner cameras ***Covid-19 applications***	**Public spaces**  *Red signs identify people too close to each other*	Use of camera-based technologies to track people's locations and guarantee social distancing. These technologies can be applied to public spaces (airports, streets, supermarkets etc.)
Thermal body scanner cameras ***Other applications***	**Retail stores**  *Red areas identify most frequented spaces in stores*	Use of camera-based technologies to analyze customers' behavior in stores by examining two main things: where the customers walk, and what items they stop to touch and pick up

AlixPartners

of supply and with a great focus on the monitoring of people's behavior, to ensure some minimum health safety standard at the workplace. Activities have included the sanitization of the workplace and the continuous testing of the employees rejoining this and their surveillance as they move from home to the workplace and back. A number of digital solutions and IoT-based tools can provide the monitoring support for all these activities and be extended across entire supply chains, as safety standard would likely be defined by their weakest links. Cy-phy tools could also help in ensuring the compliance on minimum legal requirements and protection and in managing diffusive communication with all stakeholders. They can also help in designing and managing "distributed" rather than a "centralized" business, as many thought leaders (and e.g., bosses in the city of London (Thomas 2020)) are predicting a fundamental shift in own companies will use offices in the future, with greater flexibility set to stay after the pandemic ends. This distributed approach will, according to the City Network members, remain a permanent feature of white-collar life, accelerating, for example, according to Jean Pierre Mustier, CEO of UniCredit, the shift towards "remote banking," such as advisory and call center services. Other executives at banks, including Barclays and Credit Suisse (which between them employ 12,000 people in Canary Wharf in London), have indicated that a full return to the office is likely to take some time, with obvious implications not just on the value of offices and commercial properties but also on the residential ones, with prices and sales falling (Hammond 2020).

In the restart-transition period, in fact, demand will also need to be reactivated, and for the mid-term (likely expected to be anything between 12 and 24 months for the COVID-19 or until the mass availability of a set of effective vaccines), the sustainability from a financial perspective will need to be ensured, given the high uncertainty of expected scenarios. What would be the sustainable demand over transition period? What would be the required business pick up and time to mean-reversion? And what optimal rebalance of new e-commerce activities versus the physical one could be expected? Cy-phy tools and approaches can help in addressing these points as well, providing intelligence on the basis of, necessarily, short-term memory streams of real-time data and AI analytics. Topics that have become part of most urgent workstreams for corporates includes swings in demand, not just in its size, but also for its composition; revised marketing mix and adaptive pricing; cash and working capital dynamic rebalance and targeted, transitional capital structure; flexible access to government funds, special schemes and tax shields; and the management of the financial sustainability of their entire value chain (including the promotion of "supply-chain financing," where the invoices are discounted by an explicit guarantee of the large corporate—usually better credit-rated—of the value chain or industrial cluster of reference).

Finally, as the restart and transition periods for "this" crisis will get to an end, an "other" normal and its related steady state should be targeted and designed, with a higher, targeted resilience built inside-out, with buffers and redundancies, geographical global-local diversification, alternative sets of offerings and locations, "bubble" teams and distributed workers, and much else, on the basis of a rational assessment of all the inherent and unavoidable trade-offs. For all this resetting and redesign, as corporates, families, and cities are aiming to reach the "other (sustainable) normal" ahead of them, cy-phy approaches and tools and twins (twin supply chains, twin headquarters, twin homes, twin cities, and so on) will all be of great help. Still, the challenge ahead of us will be as much psychological and physical. It will entail "new ways to make a living"—as suggested by *The Economist.* Most workers will continue to shun the office, even as schools reopen—latest data from *The Economist* suggest that only half of people in five big European countries will spend every working day in the office by the end of 2020, and a quarter will likely remain at home full-time. Given social-distance rules, no matter what the digital technology used, current offices setup could work at a capacity of 25–60% to ensure a 2 m distance between workers. And the queues at the elevators are going to be a nightmare, for high-rise buildings and skyscrapers, and the same could apply to public transport (or the use of private car that is getting back on fashion, given the residual fear of COVID–19. As implied, most of this new way of living 8 and make a living) which is likely going to be for the better, in the workspace and in social life, but not necessarily so.[5] It is a big challenge for the sustainable cities of the future to make sure this is going to be better than the past.

5.7 Creative Destruction at the Time of the Coronavirus

In a novel by Gabriel Garcia Marquez, *Love in the time of Cholera*, an apparent paradox is described and summarized in one of the author's statements on how "lovers (may) find love in their golden years—in their seventies, when death is all around them." Similarly, we could paraphrase, at the time of coronavirus, we may have found similar paradoxes—as we rediscover critical things that we thought had been forgotten by more and more digital societies and super-efficient markets. On one side, as our in-person, social activities have been mostly suspended during the parallel lockdown of over half of the global population, we have come to rediscovered the importance of the physical dimension and of its continuous data flow that is critical for its intelligence to be captured, re-elaborated, and enmeshed in the digital dimension. At the time of the COVID-19, and during the un-locking and restart phases, we have also come to understand how important it is to be able to design an almost perfect digital twin of the city and of its citizens, to allow their continuous monitoring for health and public good reasons, by capturing offline data via the most sophisticated IoT tools, 4 and 5G connectivity networks and (ideally), pooled public and private database, open to third parties and given the respect of the ownership and confidentiality principles that are already captured by the GDPR and

[5]What a way to make a living, *The Economist*, September 12, 2020.

able to develop AI algorithms for real-time diagnostic, assistance, prevention, tracing, and so on. In short—and this may be the first lesson coming from the paradox—during a pandemic, we have come to learn how digital is relevant for the physical dimension but also vice versa. The world of bites may become almost irrelevant if not grounded on the physical realm, and purely online information is of much less relevance if this is not enriched and intertwined with the one coming from the offline world. Atoms' world is instead becoming potentially unmanageable, for its too many frailties, if not supported by a digital twin helping out on its many challenges.

On the other side, a second lesson can also be derived by the experience lived at the time of the coronavirus. As the world has become more globally interconnected, with technologies ideally able to kill time and space and with the secular prevalence of urbanization and of mega cities able to attract and concentrate huge numbers of the globe's population in a few square kilometers, all of this has shown its inherent instability—leading maybe to our second paradox. This refers to the fact that secular urbanization could now be confronted by extreme tensions to get back to localization and to almost basic survival techniques. New post pandemic trends could even suggest some reverse migration towards rural areas, where people may feel safer (and live better) when locked-down, with food supplies at hand and with open spaces where social distance can be maintained and less concentration risk could be related to extreme risk factors such as viruses or pollution. According to Aldo Bonomi, there could even be a return (speaking for Italy) to its "rinascimento" model, with a country dominated by 100 cities closely interlinked with their territories, with new urban spaces and a progressive shift towards peripheral areas.[6] This would imply, for policy makers, a shift from the smart city to a smart land—where the value of immediate proximity is somehow challenged by the hyper-connectivity available that shifts some power away from city centers.

These and other paradoxes could be identified, as we live through such crucibles today of having to face a global pandemic, an existential threat posed by climate change, and the numerous economic and political trade-offs associated with them. We are emerging from the COVID-19 crisis, with a number of "creative destructions" (the essential fact about capitalism, according to Joseph Schumpeter) that are now happening as part of the process to get to one, and many, other normals. However, as a common basis for these paradoxes, we have assumed that info-telligence, as a force that shapes societies and markets, will further grow and develop. We have hypothesized that it will assume new forms and shapes and mostly within the context of sustainable cities—to dominate our ways of working and living. The result of this could be either very bad or very good—with enjoyable, democratic, and free, sustainable "cy-phy" cities or with autocratic politicians and business monopolies, where surveillance is the rule of the game and our decision-making is influenced by an artificial brain thinking in our instance, as dictated by special political or business interests. As Henry Kissinger has recently put it,

[6]Oltre la città Connessa, Massimiliano Del Barba, Corriere della sera, October 27, 2020.

"nations (and cities) cohere and flourish on the belief that their institutions can foresee calamity, arrest its impact and restore stability (...). When the Covid 19 pandemic is over, many Countries' Institutions will be perceived as having failed (...). The reality is the world will never be the same after Coronavirus." Which kind of world is for nations to tell—with countries like Estonia and Singapore that are seizing the critical moment and the available digital breakthrough to create flexible tech-enabled platforms for their citizens, according to a model they call government as a service (GaaS). And others that are using the emergency situation as an excuse to increase their autocratic, undemocratic dominance, with digital surveillance and control. The central idea of this book is that for cities to play a critical role in this leveraging technology to develop a new humanism that improves our resilience as a species, but also enhancing our overall freedom, within the spirit and the constraint of the public good. Because what a sustainable city is and stands for should be collectively defined by the augmented conscience of its citizens and not vice versa. It is, as of today, far from clear what policy making approach will emerge as a winner of the pandemic crisis and of the many others to come. This, in itself, is already a good reason to worry.

5.8 Public Good

All over the world, from Europe to the Middle East, from the Americas to Asia, all kind of governments have been keen supporters of real estate as a sector—and of cities as its largest and most sizeable and visible manifestation. Lots of public finance (e.g., some portion of the enormous wealth of the tyrant of the day, trying to buy appeasement and prevent revolts; or of the public debt financed by the citizens of the democracy of the week, looking to inflate its way out of economic troubles) have gone to support real estate. Indeed the public finance support to real estate has a long history, where banks have also had a big role, as tool of the monetary policies of the Central Bank and (or) as recipient and transmitter of the subsidies provided by the government—these has been massively obvious with the reconstruction after World war II, but ever present across some many political and election cycles, often leading to the creation (and burst) of real estate bubbles, as in the 2008 global crisis. Specifically, the main objectives pursued by public finance have been multifaceted but closely intertwined. Three, in our recollection, are mainly standing out in the international experience of the last 75 years. The first has evolved through time, to support the re-construction, development, and regeneration of main, metropolitan and regional, local cities and of large urban and suburban areas as a key driver of the countries' competitiveness and ability to grow its wealth and well-being, across geopolitical areas, and with specific reference to "special zones" of strategic relevance (e.g., critical industries, health, education, tourism, commerce). The second has instead meant to mainly support the civic commitment to the country and its social stability, closely associated with home ownership and with affordable housing (affordable rents) for the less wealthy citizens, also supporting the peaceful and positive co-existence of blue-collar, white-collar, and wealthy private individuals, given trends in cities gentrification and local and global migration.

Finally, a third objective has more recently focused on supporting the overall optimization of a key component of the wealth of the nations: real estate and the real estate (as an investable asset class) of main cities. Globally, residential real estate, as an asset class, is worth around $ 200 Tn and is much larger than the sum of all equity, bond, and commodity assets. Also, for most countries, real estate stands as the larger share of the private wealth owned by its families (also called households). Not only most people (around 75% according to several studies) are expected to live in urban areas by 2050, but also, given the different value that real estate has in urban versus suburban and rural areas, the concentration of the overall real estate wealth in global cities could be huge—and notwithstanding the pandemics. It follows that this real estate is also the main focus of the government backed public finance initiatives—across the globe. From Kuala Lumpur to Berlin, from Singapore to Montreal, internationally, we have seen a number of schemes that involved the government and public finance support to the real estate sector and with specific reference to the housing component (residential assets meant for ownership or rent by families). Those have included financial guarantees offered by the government to retail (e.g., families) or corporate (e.g., real estate developers) and real estate mortgages offered at subsidized prices and often at a loss (e.g., Fannie Mae, Freddie Mac in the United States during the last few economic cycles). Also, tax credits have been offered upon the purchase or rent of real estate assets by families or for their full refurbishment and upgrade to higher standards (e.g., Grade A, "zero carbon" certification, global safety standards, improvement, and enlargement—e.g., see the recent post COVID-19 program in Italy that is also meant to support the real estate sector and its investment and consumption cycles).

Also, in many instances, regulated pricing has been enforced on a quota share of new homes and on rents for specific neighborhoods or entire cities and even at country level, to ensure the required supply of affordable homes, given changing demographics and trends in the gentrification of cities and on the migration of less wealthy residents across countries and regions (and from rural to urban areas). As we have seen, after the COVID-19 health crisis, not only will real estate—and real estate in main cities specifically—be seen as an even more critical driver for the reconstruction and recovery of the impacted economies. But also, it will be requiring a more and more urgent radical redesign, redevelopment, or regeneration, to make it more "sustainable" at the light of future crisis—from pandemics to climate change to regional tensions and military crisis that could lead to unrest. Specifically, "smart cities" with "smart building" and "smart homes/offices" will be a key components of future investments, and governments across the globe could have an even greater interest and urgency to act to drive this transformation, supporting the private sector and without "crowding out." Many governments, in western and eastern countries alike, and regional and local municipalities, are now mostly thinking about that, but with a big constraint to face.

5.9 Finance, Real Estate, and Sustainable Cities

In another book[7] of ours, we have discussed at length the virtuous circle between (private) finance, real estate (and technology) to support the creation of competitive and attractive cities, and the creation of shared and sustainable wealth-being. The focus of this book has then shifted from the sustainability of the value created to the one of the cities themselves, at the light of the pandemic, the climate change threat, and the many more crisis they will likely need to face. Given the brutal impact of the COVID-19 crisis, a larger and larger (and sometimes worrying) role of the State and of public finance in driving the reconstruction of the economy and the development of sustainable cities is becoming more and more apparent. Assuming this public finance is supporting a private finance, market based, competitive approach, it would be welcome to accelerate this shift to sustainability. However, across the globe, governments are also running short of capital for new investments and public debt ratios have ballooned because of the lax fiscal policies pursued to limit the economic damages of coronavirus and the parallel contraction of GDP (at the denominator). In this instance, financial services, as a mix of markets, players, and products, could also provide some help and allow governments to pursue their active policies in a more efficient and effective way—e.g., potentially tapping the huge stocks of private institutional and retail savings that are still available in many developed countries as liquid financial resources. They could do this through the capital markets and with a financial institution setup (e.g., as a bank, or as an asset or wealth manager; as broker dealer or as an insurance company or pension fund). Or, at least, they could do this with the help of regulated financial products (e.g., with real estate or credit to real estate funds, securitization vehicles, financial holding structures with long-term capital, e.g., ELTIF, European Long-Term Investment Funds in the European Union, or the PIR Alternative in Italy).

Multiple options could be considered by governments to that regard and in order to pursue not only its main objective of supporting affordable, competitive, and compelling housing to their citizens but also to potentially pursue the broader range of objectives usually associated with their support to the real estate sector as highlighted above. For example, a financial institution license or the setup of specifically regulated special purpose vehicles could allow to governments to raise third party money through bonds issuance, securitization tranches, mutual fund "funnels", equity crowd funding, or even through deposit gathering. Governments could, for example, consider a wide range of options including setting up a bank that is able to gather deposits online or via physical branches (e.g., leveraging the Post Office network) or (as a wholesale bank) through certificates, term deposits and bonds, and any other money market or savings products that could be sold by retail and universal and retail banks in the country of reference, with standard agreements set at national level. They could also consider setting up an insurance company or a pension fund that can raise and manage a "float" of money (the premia paid by policy

[7]Finance, real estate and wealth-being, Claudio Scardovi, BPU - EGEA, 2020.

holders that have an inverted cash flow cycle vis-à-vis the one of real estate development or investment—with their long term mortgages), as investing in long-term real estate mortgages is a natural hedge to their typical liabilities. They could then consider to set up an investment and asset manager, able to structure a number of targeted investment vehicles and solutions (e.g., real estate multi-seller funds, equity or debt real estate funds, REIT, etc.) that could raise money from private institutions and retail investors alike and manage their savings with a view of optimizing the country's real estate overall wealth-being. In an even more ambitious and holistic vision, a strategy that develops a real estate wealth-being (wealth and well-being) manager could be also pursued with a mix of set ups and financial product offerings that could include most of the above-noted propositions and with a view of having a real impact.

No matter what the final solution, a public finance support to drive and stir the public good component of cities, and certainly including the critical infrastructure for ensuring optimal physical mobility, digital connectivity, energy usage, and more, would be critical to support a private ecosystem of investors, developers, designer, servicers, and users that would then shape the city of the future, in an optimal and sustainable way. Because no wealth-being created will be sustainable if the city is not, and we will all, collectively, have limited chances to succeed if public and private finance, the governments, and the animal spirits of the markets will not be able to find a way to collaborate synergistically on this tantamount challenge—so much more critical because of the economic and humanitarian crisis we are all living and with the climate change challenge to come. The fight over the future of offices and of the workplace may have just began, as observed by *The Economist*.[8] In fact, there is so much more at stake, for real estate and for the future of cities, with—for the governments—the temptation to turn the clock back to limit the economic damages, from the collapse of the city-center cafes to the USD 16 Bn budget shortfall that (for example) New York's subway system is facing. But rather than resisting the brutal forces of change now in action, just accelerated by COVID-19 as, as we have argued, digital change and many other threats (climate change as likely the most critical) are eventually the real driving forces behind all this. Instead to offer subsidies to sustain zombie sectors and promote initiatives aimed at "change everything not to change anything," it is more promising to anticipate the consequences to come and act, through redesign, regeneration, redevelopment, and the overall transformation, with sustainability as a target. For all this, cities are the top priority, with city centers obviously standing out, but not only. New complex urban-planning rules will need urgent simplification to attract private capital and promote collaboration with market forces, not resistance. Urban policies will then need a radical overhaul to allow buildings and entire districts and even cities to be redeveloped for new uses, including offices but also flats, manufacturing, and entertainment. This is no small task for governments across the globe: it will require the subtle and smart working of

[8]Office politics, *The Economist*, September 12, 2020.

human brain cells more than the brutal and inept use of the capital provided by taxpayers, public debt, or else.

References

George Hammond, Canary Wharf has a flood of new homes but only a trickle of returning workers, Financial Times, September 16th, 2020
Mari Hui, Hong Kong is using tracker wristbands to geofence people under coronavirus quarantine, Quartz, March 20th, 2020
Daniel Susskind, Offices have a future – but what about other workplaces?, October 19th, 2020
Daniel Thomas, Bosses predict permanent shift in working and an evolution for cities, Financial Times, September 17th, 2020

Chapter 6
The Rise of the Cy-Phy Company

6.1 Smart, Cyborg, and Singularity Driven

Most global industries have been impacted in the last few decades by huge value migrations. It would be hard to put a precise date on when it all started (creative destruction is, after all, a continuous process and the very engine of growth for the capitalistic model). It would be even more difficult to put clear references and well-defined assumptions on the main change agents and drivers of these—digital innovation (a "misnomer" per se) being likely one of the most often cited factors of transformation (transformation being also a "misnomer"). However, it is safe to postulate that the pace of change has been consistently accelerating, with deeper and wider disruptions happening more often and in unexpected ways. The very notion of "tail risk" is now under scrutiny—with so many "black swans" flying around, it is difficult to see the linearity element in all this, not to mention the normality of a distribution that is based on large numbers and predictable casualty. In this context of extreme and accelerated, multi-faceted disruptions, "smart cities" have been ailed and positioned as a good force for change. Hosting more than half of the global population and set to rise further, they have been promising "smarter" ways of doing things—basically, leveraging digital technology to offer better use cases on such utilities such as energy consumption, mobility, and safety. The initial hype linked to large development and urban regeneration programs designed around these use cases has so far fallen short of expectations—with real estate values and the very attractiveness of the city in question failing to capture the value arising from the "smart city" component of these. In answer to this, a number of new technological advances are already being set in place—from 5G (mobile connectivity, with real-time and enhanced, localized hedge computing) to IoT (always charged, smaller, and even implantable) to multi-cloud-based solutions where AI/ML apps are becoming a core part of the software as a service proposition. As a result, the "smart" city is becoming "cyber," with greater centrality put on the individual and on the specific services he may need, for business or leisure. Digital interaction is augmented by more

© The Author(s), under exclusive license to Springer Nature Switzerland AG 2021

C. Scardovi, *Sustainable Cities*, https://doi.org/10.1007/978-3-030-68438-9_6

personalized and real-time feedbacks, with services dynamically adjusting themselves (hence cybernetic).

Alas, cyber is not enough if the physical dimension is not, in parallel, also augmented, by capturing its offline data flows and mining their inherent intelligence. In turn, and more importantly, this intelligence needs to be embedded into the cyber dimension (the digital twin) and, vice versa, leveraging online data to embed new intelligence in the physical dimension. The result of this fusion is much more than a more effective and efficient (a more productive) city—as resilience is built inside out, ensuring a greater sustainability for the city and its ecosystem. Sustainability is in fact something more than mere resilience—it stands for the capacity of the city to survive through near extinction events and all kind of health, financial, climate, geopolitical-related crisis, but also for ensuring that the rule of freedom and democracy are kept, with a compassionate social focus to extended inclusion, for the preservation of wealth and well-being. In this sustainable setup, we will all feel more comfortable in our evolutionary process to become more and more android-like, cyborg even, with implanted IoT and other high-tech devices (including exoskeleton able to augment our physical prowess and address certain handicaps and illness), with the required safety standards—related not as much to the Internet, but to its billions of end points that are more vulnerable to hacking and cyber thefts. If a proper, seamless setup able to integrate the digital and physical dimensions, with golden standards of cyber security, is set up, as individuals we will be in a position of benefiting from the progressive convergence of our digital and physical dimensions—we will think smarter and move faster, augmenting our sphere of influence and cognitive and execution powers but always on the basis of our conscience and internal set of beliefs. In a sustainable city, androids and cyborgs will still be a humanism—made around and for the human species—to help this in becoming more resilient, stronger, and better. In this more sustainable setup, winning companies and institutions of all kind—public and private—will then flourish because of their abilities in driving this cyber-physical approach to new heights and then building singularities in their business models based on the aggregation and analysis of data coming from both the online and offline worlds. Robotization procedures would then extend the potential impacts and reach of these singularity explosions—similarly to what high-tech, prostatic artifacts could do with android-like people workers. In their search for an optimal and more sustainable way for the digital and physical dimensions to coexist and positively evolve—as a central anchor of a new human way of living and more closely intertwined—they will likely revolutionize (again) most of the main industries in the years to come. In the following paragraphs, few strategic forays are depicting how financial services could change, then TMT and the telecom industry specifically and finally the healthcare sector.

6.2 The Advent of the Cy-Phy Bank

Banks have been designed and developed on a simple trick—buy the use of stable, safe money at a low interest rate and sell money on a more volatile, risky basis and at a higher rate. Acting as money brokers, they have then started playing as money conveyor-belts, managers, and advisors (when offering payments services and then asset management and private banking) and insurer even (when hedging any kind of risk with a potentially negative financial impact). They have added few other services in between, extending their middle-agent role across people, corporates, and governments and doing intermediation through time and space. Often, they have used the alchemy of financial structuring and risk management to transform lead into gold (in the short run), often with disastrous consequences in the long term (this financial innovation solutions being often "smart"—here in its worst possible meaning—but short on resilience, hence little sustainable). In our first "cy-phy" strategic foray, we consider, in the context of the city of the future, fully digital and built of huge amount of data and information, how banks could in fact become a key player of the info-telligence industry, helping to make this more resilient, transparent and fair, inclusive, and respectful of the environment—hence sustainable. From a business based on physical and (more and more so) digital money (and risk, as the key production factor of the financial services and the unavoidable nemesis of undeserved good fortune and reward), cy-phy banks could develop a new business based on information and intelligence—also immaterial as money is and acting on both the on and offline dimensions. For a start, they could offer to clients a safe way to store their data, given their own high-end cyber and physical security standards. It does not matter that, in the case of data, most banks will then rely on third party cloud partners, such as Amazon Web Services, Microsoft Azure, or Google Cloud, as they would be the contractual counterpart and first port of call. Being regulated and deeply supervised by domestic and international central banks (and, hence, ultimately by policy makers) could also ease some nerve, with regard to the ultimate scope and role they would play with such confidential and highly relevant data. And having substantial capital buffers should also help, in the way of offering financial reassurance, should anything go wrong (on the contrary, big cloud providers are both lightly and not regulated at all and loosely supervised and have no required minimum capital that they have to set aside against potential operational risks).

Secondly, banks could start advising customers on when and for which T&C (Terms and Conditions) and for what kind of fair price they should consider monetizing or capitalizing and preserving their more routine-based data and more critical information and then derived intelligence. Banks could also support them in having a thorough audit of the bidding counterpart and with a full look through on the ultimate use cases that will be pursued with their data and information. They could make them more knowledgeable and an active part of the data trade and intelligence engineering for their own benefits—supporting them from origination to valuation, from sale to post sale and with continuous monitoring of the end use cases being created. Ultimately, they could act as "consumer's advocate" and "holistic wealth manager." Assuming wealth in the future will be more and more generated by data and captured through AI, then wealth management will have an increasingly important digital dimension and a new kind of "wealth-well-being manager" should be required to manage information and intelligence, and not (just) money. Potentially, banks could extend their reach and interconnectedness across a number of retail, corporate, and institutional counterparts, extending their reach across time (as data can be stored forever and data history is also relevant for analysis) and space (data can virtually travel across boundaries with close to zero incremental costs). On this basis and given a relevant governance and a clear set of rules to ensure fairness, proper behavior, transparency, and liquidity, they could play a key role in supporting the creation of a more efficient, effective, and inclusive information and intelligence (info-telligence)-based marketplace. This marketplace would certainly need to address the current reputational issues faced by big digital players, not to mention the increasing scrutiny paid by regulators and policy makers. It would then need to ensure high cyber security standards and protocols—from storage to end points and uses.

6.3 Singularly Cyber-Physical Bank

Extending the "five pillars" framework of analysis that we used to define the "synapsis bank" (Scardovi 2018), financial institutions could capture and gather structured and unstructured data and information as their main production factor (first pillar). They could then create intelligence by leveraging applied analytics as more serviceable and valuable end product (second pillar). They would then start trading on the basis of a more sophisticated intermediation model, ideally on a many to many setup that is connecting multiple stakeholders to ensure that the best intelligence gets used in the best and most possibly valuable way—using pricing as a tool to measure value but also with the support of a well-defined framework of guidelines and principles (third pillar). Actually, more than broker-dealers of info-telligence, banks could gain back some of their competitiveness and critical relevance by providing incremental value to their ecosystem of reference if they are able to design innovative solutions (forth pillar). For these innovative solutions to be compelling and attractive, they would need to produce and use info-telligence to address relevant issues and crack hard problems, hence creating real incremental wealth and not just migrating the one already available—and mostly using AI, hence progressively becoming "singularity banks." The bank could do all this on the basis of its reputation and superior trust in the overall system, mostly digital—hence managing carefully its cyber trust and related cyber capital at risk—but not only (data coming from the offline world would then be digitized, hence requiring, on top of physical security, also some form of cyber security). In fact, extending our previous reasoning on the "singularity bank" (Scardovi 2019), a critical competitive advantage would come from its ability to use AI analytics to convert information into valuable and serviceable intelligence that can be used for specific purposes or radically new solutions.

These "intelligence explosions" could be put at work for the bank itself or sold to clients as part of an "info-telligence" proposition that becomes a proper service offering. Clients could buy raw material (data), in a logically organized way (information) or elaborated into usable and generic forms (generic intelligence on specific phenomena) or into tailored setup that are meant to address specific questions. There is in fact greater value in matching relevant intelligence with critical needs than in just producing huge brainpower with no immediate application or impact. The overall design of a new info-telligence proposition could become a potentially game-changing opportunity for a reputable, well-regulated, and supervised "singularity bank" that can actually "bank" on data that customers are willing to deposit, store, and put at use on its own fortress-safe balance sheet and IT platform (whether mostly in the cloud or not is of little relevance). This is a significant change in the banking paradigm, but with lots of continuities as well. Money is substituted by data (another intangible) as the basic mean of the trade, with the production of intelligence (as opposed to financial returns) as the main production factor and with cyber-risk becoming the nemesis of financial risks. In a first stage, the independent bank could offer storage and valuation services to data and information owners. Customers would receive a fair-value appraisal on their past data and future stream of information. Similarly to their "traditional" deposit gathering business, banks could then get a monetary yield on the stored data, allowing the bank to monetize them according to certain, restricted use cases and with greater bargaining power (and lots of lawyers keen to negotiate terms and conditions). Data, once aggregated across multiple clients could also become more valuable and get transformed into intelligence for end vendors to buy (similarly to the credit business, where money is bought, stored, manipulated, and then sold). The net yield would compensate the bank for their role as broker-dealers but also as intelligence providers.

6.4 A Digital Wealth-Being Manager

Finally, the whole evolution of the bank into a cyber-physical bank would stand up if, considering also the last of our "five pillars" (Scardovi 2018), the bank itself were to command a "security premium"—something not only related to its physical and cyber security, when storing and managing relevant data and information, but also to its "moral compass," when deciding the proper use of these and of related, derived intelligence. The fact that banks have been managing reputation as their most relevant of intangible assets and have been subject to the public scrutiny and to the supervision of regulators and policy makers—with global standards to match—should also provide some level of comfort to the end users (the private citizens, first and foremost). Being credible as a "safe and secure" storage harbor but also as a "moral and ethical" manager of our data is a very important driver to secure a relevant competitive positioning in this growing information economy. It is certainly a "qualifier," but it may not be a "winner" as well, if synapsis banks were not able to show and demonstrate that they are also becoming "singularity banks"—i.e., becoming the best in extracting serviceable truths and putting at productive work, in the best interest of their client, the intelligence derived from data. In truth, the bank itself may not be able by itself to become the best AI provider to extract truths and intelligence from data; but it could be a very good clients' advocate I selecting the best counterparts for them should, hence potentially becoming an holistic and digital "wealth-being" manager—keen to manage the monetary value that we can extract from our data, but also to optimize the well-being we can derive from them, for us and our families, sustainably.

This data wealth-being manager could advise us not just on the market and fair price we could get for our data, but also on how we could market them at best and more competitively and according to specific and superior use cases. For these, this data wealth-being manager could then advise on the best applied analytics and intelligence generating AI tools to consider—not dissimilarly from what a wealth manager does with private money. It could also offer a (then re-insured) insurance on the data traded (e.g., for theft or fraudulent use). Such data wealth-being manager could be designed and born as a dedicated company and as a fully licensed and regulated entity, with minimum required regulatory equity capital and best practice IT capabilities on applied analytics and AI/ML intelligence generation from data, with the access to the safest third party cloud-based infrastructures and with optimal internal processes and organizational setup. This data wealth-manager should also pursue a transparent governance and with the highest reputation and ethical standards to match. It would pursue a full inclusiveness approach, in collaboration with consumers' association, financial services, and data privacy/protection regulators. And it would promote an open data approach, running across all kind of data categories and sectors as opposed to the financial services-only approach currently pursued by PSD2. It could include, for example, data gathered from API that the consumers are willing to push on social website and communities and other past and future social and business data that customers are willing to trade for a given use. It could include as well other web-based data captured by public sources to mix them with private data and information in the AI-driven transformation process (e.g., traffic and weather forecast, pollution and social life/leisure, macro and micro business trends, observable prices, and so on).

This data, wealth-being manager could also include other IoT-based information that the customers are fine to supply (e.g., on their driving style via a black box that is embedded in their car, on their health style via a digital bracelet, on their home/office life—on energy consumption and safety). It could also extend and include a further set of financial/structured data that the parent bank already owns, not for mere reselling, but as a key input to derive even better competitive intelligence and value adding solutions. The promotion of further data gathering could also be sponsored by the parent bank with, for example, free current account for customers sharing their social data and a monthly payment for accessing their fit-bite or smart-home information flow. Taking into full consideration the personal and confidential nature of data, even the availability of information at cohort level could be of value, as some applied analytics on this could generate valuable views for vendor-specific use cases, at cluster and segmentational level and cut short their R&D and marketing cycles. Specific initiatives of "data crowd funding" could also be launched, to gather specific set of information, leveraging the value of the critical pool of data and shared applied analytics to generate a financial return on a specific solution. For example, the use of contact lens for people suffering of diabetes would allow the flow, capture, and mining of information on their health and lifestyle and on their multiple interconnections during their lives. This information could then be mined and analyzed and sold to big pharma, hospitals, and first aid centers and diet and training specialists for a fee or to get a more profiled, real-time health check and following monitoring and medical advice. Eventually, such a data wealth-being manager would end up doing what banks have been doing for centuries. Using data and information instead of money, buy and selling, after some more or less sophisticated repackaging, transforming something very intangible and of no obvious use (money, data) into something else more valuable and for productive uses (yield, intelligence).

This would require the ability for the bank to convince the customers to trade their data in a safer, more ethical, and rewarding way. It should also help in selling the idea that customers need financial advice and insurance protection on data, as we believe should be the case. The bank should also be able, as in its deposit-lending business, to operate multiple transformations: from low-risk, short duration, but behaviorally stable data gathered from multiple providers to high-risk, long duration data and intelligence fewer users. All kind of fundamental functions now operated by the global financial system could then be replicated by using data as opposed to money, and intelligence would be managed instead of a given risk/return target profile: from the payments and exchange of information to the setup of funds investing in data and AI-based algorithms, specialized lending guaranteed by information, and so on. On top of the traditional broker-dealer role played by banks in financial services, a data wealth-being manager would also create value by structuring and analyzing, through specific applied analytics and AI/ML, the data gathered and in answers to the use cases it could define and that the vendors are keen to address. Ultimately, the liability side of the business would be made of retail clients and the asset one by corporates keen to use the retail data to sell existing products and innovate and develop better ones. On top of the net yield on the data/intelligence traded (interest margin) and of the commission income gained for the advisory services (service margin), the data wealth-being manager could then retain, own, and capitalize in its goodwill the AI-based algorithms internally developed—it could then use them independently from the data owners and borrowers to design and execute new solutions and to pursue other entrepreneurial activities in financial services or elsewhere. A large universal bank would be likely best positioned to design and launch such a holistic data wealth manager, potentially making this a marketable and independent entity in the mid-term, with institutional investors buying into it for the long term and with sustainability in mind. But other fintech could also play the trick, moving faster and smarter.

6.5 A Data Bank Is Already Here (and Is Worth Billions)

A bank capturing data and making an income by selling them, or trading on information to make a living and even offering their value adding transformation (and valorization) into intelligence as "digital wealth-being" manager and successfully creating value for its shareholders, could all look as a very far-fetched strategic visioning and very unlikely to come by, at least in the mid-term. In fact, a data bank progressively becoming an integrated financial player which basically gathers, intermediates, transforms, and sells information and intelligence is already here and worth billions. This is Ant, one of the largest fintech globally and with a prospected listing in Hong Kong and Shanghai (leaving aside the regulatory process that is now delaying this, likely on political grounds more than on systemic stability ones[1]). When floated, Ant will likely have a much higher value than JP Morgan Chase, the world biggest bank, and with equity capital that is a high multiple of the one of Ant.[2,3] With a potential forward price-to-earnings multiple of 40, in line with big global payments companies, Ant could fetch a market capitalization of over $ 300 Bn and become more valuable than any financial institutions in the world (the most valued financial servicecs company as of today is VISA, with a market cap of around that figure, also a payments company, working on data and information and with little balance sheet and global presence). Differently from VISA, Ant is a most integrated fintech platform, offering payments but also a $30 Bn lending platform and insurance, asset management, and brokerage on the go, and with e-commerce one click away, as shown in Tables 6.1 and 6.2, that is describing Ant's developing and encompassing integrated offering and its aggressive growth in users and in the uses cases offered via an seamless customers' journey.

Spun off from Alibaba, an e-commerce firm with over 1 Bn users and that built its vast fortunes in the digital data business (as e-commerce websites are basically capturing an intermediation fee by matching suppliers with clients), Ant has got to manage over $ 16 Trn of transaction in 2019 and connecting 80 Mn merchants. As for many other "fintech," the most obvious port of entry to financial services has been payments. Leveraging data, capturing information, and creating serviceable intelligence out of it, Ant has then been able to add a lot and become more of a bank (even if, technically, it has not got a full-fledged banking license). It sells now loans to its consumers and over 6000 kinds of different health insurance products, not to mention money market funds and other asset management stuff. As many other "super apps," it has been banking on data to develop a new, lighter way of banking. Nubank and Mercado Libre (owner of Mercado Pago) are pursuing similar strategies in Latin America. And the same would be for Asia Grab and Gojek in Southeast Asia

[1]China's banking regulator signals tougher fintech antitrust laws, Sun Yu, *Financial Times*, September 11, 2020

[2]On the march, *The Economist*, September 12, 2020

[3]Beijing says it halted $37 Bn Ant IPO to protect market stability, FT reporters, *Financial Times*, November 4, 2020

Table 6.1 At its core, Ant Group is a tech company leveraging its intelligent platform to build networks and leap into new domains

蚂蚁金服 ANT FINANCIAL

Snapshot of Ant Group's Ecosystem

Payment Services & Common Platform	Wealth Management
支付宝 ALIPAY • Online payment platform • **1.2bn+** annual active users • **175m+** avg. daily transactions • **3x PayPal's** AAUs and annual transaction volumes	余额宝 蚂蚁财富 Ant Fortune • Yu'eBao online market fund • Offers higher rates than banks • **150m+** annual active users • **$157bn+** AUM • Ranked #3 globally
Banking	**Credit & Insurance**
网商银行 MYbank • Online-only banking platform • **Affordable loans** for small and micro businesses, e-commerce platform sellers • **$290bn loans** since launch in June 2015 (1% default rate)	蚂蚁小贷 ANT MICRO LOAN 芝麻信用 ZHIMA CREDIT • **SME** and **micro-loans** ($64bn) • **Insurance** products • Online credit reference/scoring • **3m cumulative users** for SME loans and 380m for insurance

Innovative technologies (face-recognition, cloud computing, AI)

Local & global partnerships

Big, dynamic data circle
(generated from the business & feed back into)

Business model

- Online payments and **financial services affiliate of** the Chinese e-commerce giant **Alibaba**
- Comprises of a wide range of services from **online payments (Alipay)** and **market fund (Yu'eBao)** to one of China's first online-only banks **MYBank** and micro-loan provider Ant Micro Loan. The diverse offering means that 80% of customers use 3 or more of their services.
- All services connect to the **AliPay App** which serves as the common platform
- Aims to provide "**inclusive financial services**" and **serves SMEs, micro enterprises** and small consumers
- It is estimated that **c.58% of China's online payment** transactions go through Alipay

Core competencies & differentiators

- Develops a **collaborative environment** among customers, suppliers and partners and **creates network effects** based on huge customer base
- **Leverages** the power of **Alibaba's e-commerce business** and **big data obtained** across Alibaba eco-system (re-invests data generated back to the businesses)
- **Penetrates into various consumption scenarios** in people's daily life (from shopping and gaming to transportation, dining and financial services needs)
- **Incubates innovative financial products**, including building up its own financial cloud technology platform

Source: Ant Group website; desktop research and AlixPartners analysis

AlixPartners

Table 6.2 Ant Group aims to increase its active customer base globally to 2 Bn by 2025 with a particular focus on foreign markets

ANT CASE STUDY

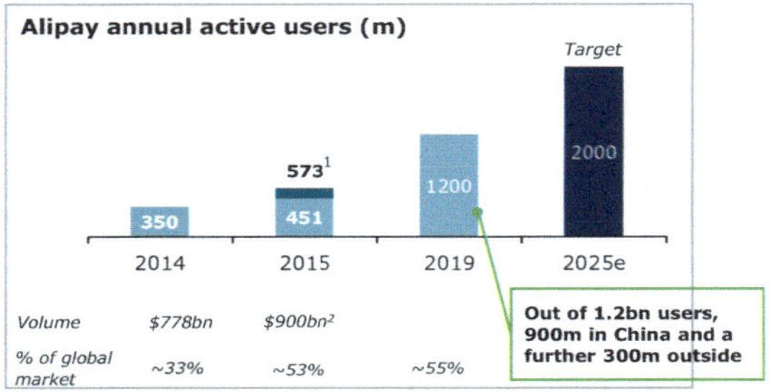

Strategies for the future

- **Expand globally into the US** and penetrate further to the **EMEA markets**
- Serve not just Chinese tourists visiting foreign countries but **penetrate into local markets / consumers**
- Reach to **2bn customers by 2025** globally and gain a bigger portion of the global mobile payments market which is estimated to reach $3tn by 2021
- Use **technology as a key strategic enabler** to drive the growth and expansion plans

Ant Group's platform creating rich customer data and gaining network effects as customer base increases

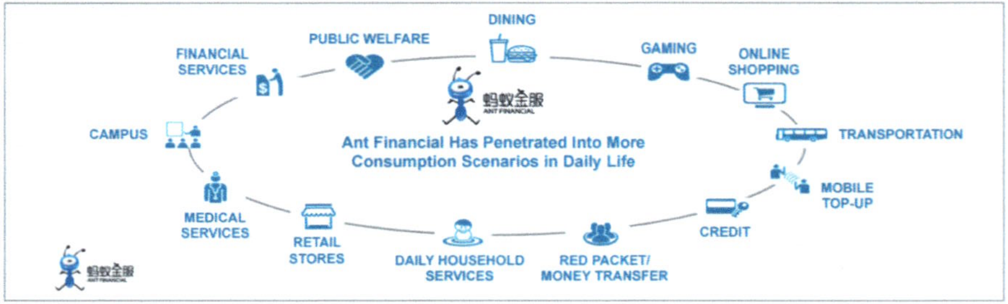

1 Including Paytm users (Ant Financial invested in India's Paytm)
Source: PYMNTS.com; CS report from Feb 16., desktop research and AlixPartners analysis

AlixPartners

(starting from their ride-hailing services instead). Data is in fact the key differentiator of their business model. Ant, for example, masters more than 3000 variables into its credit-risk models, and its automated systems decide whether to grant loans within 3 min—it does then sell loans coming from third parties, mostly traditional banks, on the basis of its track record in getting this right. It is then banking on data and AI to extend banking products and without actually bearing the brunt of being a bank, with a balance sheet, a strict regulatory regime, and large capital requirements. Ant is a data-bank also in the way it is able to use information to offer solutions on the spot that then leads to better customers' experience and more cross-sell and fidelization.

For example, Huabei, a revolving, unsecured "virtual" credit card, is offered at the tap of a finger to clients that are considering an e-commerce transaction and looking for (at the last second) a deferred payment by a month or by installments. Then Jiebei, also a feature of Ant, can offer larger sums and loans as well, for retail and small businesses, with a hefty interest rate. Also in this case, the trade is more on the information than on the money itself, as Ant owns a huge data set and can create superior intelligence on the creditworthiness of the borrowers, based on the knowledge of what they do online—when buying, selling, or looking around. Money is eventually the commodity part of the value chain that can be offered by Ant itself or by any other direct lender or traditional bank. Eventually, after a first phase of "origination to repackage and sell," Ant is just assessing borrowers for a "technology fee" and then passing them on to banks that extend the loan (and ultimately bear all of its risks). Ant retains then a cash-rich, capital light model banking on data and selling information and intelligence and adding value the more data are generated by these loans and based on their final outcome. Because of its business in asset management and insurance, Ant could also be considered a nice proxy of the "digital wealth-being" bank to come. In fact, it got started in asset management with Yu'ebao ("leftover treasure"), with merchants and shoppers leaving small sums at each purchase for parking and investing in a money market fund (given the number of counterparts involved, this lead to the rise of the largest money market fund by 2017).[4]

It has since become one of the biggest distribution channels for investments in China, covering both equities and bonds, again acting as a data-information bank—as for the lending business, Ant screens perspective clients and directs them to products (managed by others) and collecting again a service fee. From asset management to insurance, the step has been small, with even greater value to be gotten from data as the health of a perspective customers can be gauged on the basis of his or her digital activity, including how it eats, what products it is looking for, for how long, and at what time of the day. The more data will be available for each and every single business, and on top of the e-commerce data of the parent company, the greater, for the network effect, the value of the data, information, and producible intelligence that Ant will be able to capture and monetize for business purposes. It could be a safe bet that, being able to aggregate e-commerce, social and gaming data,

[4]Queen of the colony, *The Economist*, October 12, 2020

with financial data, Ant and its parent company Alibaba could progressively be able to understand better and better the profile of their customers, anticipating their needs and proposing them relevant solutions, influencing their view and contributing to inform (and form?) their decisions. Should it become able to aggregate also offline data, from an IoT network embedded in a cyber building, distributed in digital neighborhoods, designed inside out in smart cities, it could become an almost unbeatable force, banking on data and AI, to know everything about ourselves, and more. A reason more to worry about from its Chinese policy makers and regulators, for endgames unknown.

6.6 A Star Is Born

For all its beauty and incredible commercial and financial success, Ant is by no means the ideal model of what an innovative, truly big data and AI-based, "cy-phy" bank could look like. Of course, it would be easier to do even better to start from such a larger market as China, where sensibilities and constraints to the use of data are much more limited than in, say, the GDPR-ruled European Union. However, it would be highly hypothetical, but nonetheless plausible, assume that such a "star" could be born in Italy, of all places, and based on an innovative combination of cyber and physical assets. As the knowledge economy becomes the norm and the foundation of a data and AI-based economy, big data platform aggregators could become more and more valuable because of the ownership of data, applied analytics, and AI algorithms, we have argued. And they would use the resulting intelligence to sustain their own, newly designed, innovative solutions—potentially applicable across all sectors and starting from banking, the mother of all brokerage-based businesses. This is likely a "winner-takes-it-all" competitive battlefield and where no one acts right now as "consumer's advocate," advising customers on when, which T&C, and for which fair price they should consider monetizing their data. They could support them in having a full look through on the ultimate use cases that will be pursued with their data. They could make them more knowledgeable and an active part of the data trade and intelligence engineering for their own benefits. This role of "consumer's advocate," also supporting the creation of a more efficient, effective, and inclusive information and intelligence marketplace would need to address the current reputational issues faced by big digital players, not to mention the increased scrutiny paid by regulators and policy makers, keen to ensure high cyber security standards and in full compliance with laws and regulations (but also to influence their citizens in non-democratic settings?). This is a potentially game-changing opportunity for a reputable "singularity bank" that "banks" on data that clients are willing to deposit, store, and put at use on its fortress-safe balance sheet and IT platform.

This is a significant change in the banking paradigm, but with lots of continuities as well. Money is substituted by data as the basic mean of the trade with intelligence (as opposed to risk return optimization) as the main production factor. Customers receive a fair market value independent valuation based on their past data and future stream of information based on AI apps. And, similarly to the deposit gathering business, they can then get a monetary yield for that, also calculated by AI apps matching potential supply and demand and allowing the bank to use them for certain use cases once aggregated and transformed into intelligence for end vendors to pay (similarly to the credit business). The net yield would compensate the bank for their role as broker but also as intelligence provider. Alternatively, the bank can advise on the best use of the customers data and market them competitively and according to specific use cases (and after applied analytics and intelligence generation by AI apps), similarly to the asset and wealth management business. It could also offer (re-insured) insurance on the data traded (e.g., to cover their fraudulent use) also underwritten on the basis of AI algorithms, gauging risks, and related expected costs. Such a "singularity bank" could include data gathered from API that the consumers are willing to push on social website and communities and other past and future social and business data that the customers are willing to trade and for a given use. It could also include other web-based data captured by public sources to mix to private data in its AI-driven transformation value chain (e.g., traffic and weather, pollution and social life and leisure, macro and micro business trends). Opportunities would be limitless. This singularity bank could also include other IoT-based data that the customers are willing to supply, for example, on their driving style via a black box, on their health style via a digital bracelet, and on their home and office life—on energy consumption, safety, and so on. It could also include a set of financial and structured data that the parent banks already owns (assuming it is born out of a traditional player, or in alliance with a few of them): not for mere reselling, but as a key input to non-financial services competitive intelligence transformation to sell to non-FS vendors. Specific initiatives of "data crowd funding" could also be launched, to gather large sets of information, leveraging the value of the critical pool of data and shared applied analytics, to generate a financial return or a specific solution. For example, the use of contact lens for diabetic people could be promoted and financed for a song, to support the gathering and selling of data and intelligence to big pharma, hospitals and first aid centers, and diet and training specialists, for a financial return for the customer, or to get to a profiled, real-time, medical advice, with the singularity bank acting in between, to design and trade, ensure, and provide safe solutions—helping our health in the process.

A specific application would then be of particular interest and relevance. Through crowd funding, leveraging the data and information coming from retail investors of all kind, and others coming from the real economy—corporates and cities—a star singularity bank could match the supply of spare money to the demand of long-term, illiquid investments that are now required to recapitalize and transform, with real equity and not with further debt (whether private or public), our countries after the pandemics. A cy-phy bank would be better placed to do just that, funneling a small share of the financial savings of families and individuals towards the best opportunity to create sustainable wealth, supporting the conversion to green, the adoption of innovation, and the transition to an AI-dominated economy. It would support the creation of an inclusive "capitalism for everybody," thanks to a seamless process of communication and information between the digital and physical realms. In doing so, it would end up fulfilling the highest mission of a truly vocational wealth-being manager, helping people to put their money (and digital wallet) where their heart is, supporting the convergence between their investment strategies and civic conscience and behavior as citizens and as an active contributors to the success of their city—hence of themselves. This allocation of scarce resources and of even scarcer opportunities would then pursue financial and well-being success.

6.7 From Dumb Money to Dumb Cables

Similarly to banks and to other financial services, traditional TMT (telecom media and technology) have lost already a few competitive battles vis-à-vis the "big digitals" that are now dominating the most valuable markets. Competitive battles of the past, between content and context (the content provided by media and the context provided by web portals and all the hardware and infrastructure provided in between by tech companies), have already been superseded by others, relegating a few sectors (and many more players) to irrelevance. New fights are started, at technological level, to pursue the latest innovation and the highest capacity of connectivity, or calculation, for example. A most relevant and recent example is the competitive battle fought for the 5G, with all the geopolitical sensitivities and implications that are making this competition even foggier. For the sake of this discussion on smart and sustainable cities, our starting assumption is that if telecoms (and TMT more broadly) will fight this game at infrastructure-level only, with huge investments on 5G new technology, they could end up just as carriers of others' value migrations to come. A complementary and augmented proposition for them to consider would instead consider developing in the data finding, capturing, and mining space, to drive singularities and design intelligence-based solutions—the real new "content" of the future. The idea would then be to leverage the telecom relevance on online traffic and physical presence (through technical equipment and shops) to aggregate data and create intelligence via AI applications that are aimed at designing new propositions for "smart-cities" and "cyborg-like" citizens, for business and leisure pursuits. Telecoms, as owners of cable and of the new 5G technology, may have left eventually some remaining first mover advantage and could leverage this to get smarter in data acquisition and aggregation, offering, for example, connectivity at a discount in exchange of data, IoT for free in exchange for the right of use of captured (and anonymized) information, and so on—trading information on trick, as done in the past by most "big digitals," but also offering solutions as end product of the gained intelligence.

Telecom companies, while offering connectivity and trying to aggregate data to test and learn and create, independently or jointly with other counterparts, valuable AI apps to address the hard problems of cities, could also become the introducer and natural "conveyor belt" of new technologies that can be now facilitated by data and powered by AI. In fact, a long list of this is already well-known and even likely to be adopted by 2022[5] by companies. They include user and entity big data analytics, application and web-enabled markets, Internet of Things, machine learning for routine processes and robotization, cloud computing, digital trade, augmented and virtual reality, encryption, new and senseable materials, wearable electronics, block chain and distributed ledger, 3D printing, autonomous transport, quantum computing, non-humanoid land robots, biotechnology, humanoid robots, aerial and underwater robots, and much else to follow. If this is the hard content that telecoms could support and help in democratize and distribute, a great work of design and fine-tuning would then also be required, to make sure all these big innovations come to fruition as something relevant, serviceable, critical for solving the needs of cities and, most importantly, of their citizens—putting back the technology at the service of the people at the center. Eventually, most of these technologies are much less relevant if there is dodgy connectivity and become easier to adopt if embedded in the physical space of the city, from squares and streets to buildings and home/office fruition, through IoT—also usually produced and commercialized by telecom companies. When ready and made available at scale, if well designed and implemented, all these AI-powered new technologies could dramatically revolutionize businesses and societies. Big digitals, while big time "most likely winner" in this new cy-phy new, all-out competition game), would not necessarily need to end up as such. A "data bank" could, for example, have a chance on this. As could telecoms also be, with all they physical (cables and antennas, shops, and networks of technicians). As real estate developers could be as well—leveraging land, bricks, bridges, and tarmac as starting point.

6.8 Cy-Phy in the TMT Industry

The all world, starting from its main cities, we conceptualized at the beginning of this book and in a (now bygone) pre-COVID-19 era, is becoming more and more "cy-phy." A "cy-phy" world with bites and atoms seamlessly integrating and complementing each other to reproduce almost perfect "twin" ecosystems, with the physical feeding into the cyber one and vice versa. Not only this cyber-physical "fusion" is providing new and augmented ways of working and living in smart, cybernetic, and even "meta" cities, but, we discussed, it should support more resilience as well and a more balanced approach to the management of the ecosystem—to make the city "sustainable" and not just "smart." That vision still holds true, in a post COVID-19 world, but subject to a dramatic acceleration in place, as most countries are trying to unlock and restart—repairing wrecked supply chains and reactivating demand and with a greater and expanded role required for telecom companies to support hyperconnectivity and almost real-time 5G-based services.

[5]The impact of AI on business and society, Lucy Colback, *Financial Times*, October 15

This acceleration comes from two main reasons. On one side, because of the crisis, the adoption (and cultural acceptance) of continuous monitoring through all kind of wearables and location-based IoT devices has taken off with the sponsorship of public institutions, influencers, and (in some cases) the enforcement by the rule of law or by the brutal forces of dictatorial regimes and leveraging on the fear, or sense of duty, of companies and private citizens. If we are now all monitored for body temperature checks, social distancing, and contact tracing (or just to check if we wear the required protections), we all feel safer, or—if we don't fear the virus for ourselves—we feel we contribute to the safety of the weaker people that are more at risk and to the sustainability of our health system.

On the other side, as a direct consequence of this, the overall technological infrastructure has been speeded up, with IoT now commonly placed in public, open spaces and on offer for quick installation and use for shops, offices, plants, and school, but also at household level. This acceleration may have been driven by the aim to slow and possibly stop the spread of the pandemic, but—once in place—it has been putting the basis for much more sophisticated and mid-long-term uses. In fact, demand activation could mean, for example, not only the promotion of products and services on offer to address short-term needs. But also, it could mean the gathering of off- and online data to understand how our behavior has changed as a consequence of the lockdown. Likely, nothing will look like before, as we work and live in different ways, with different concepts of mobility, sociality, and well-being. This challenge and the related opportunities are specifically paramount for the telecom industry as it becomes obvious how competing on the fiber and cable networks or on the 5G next generation of mobile technology is not enough. In the other normal(s), we are trying to reach after the pandemic, and given the climate change challenge ahead, we will all need to become more "cy-phy" and leveraging big data and AI to their full maturity. For "cy-phy" telecom players, this could mean addressing three emerging disruptions. The first one, related to the dynamic unlocking (and potential relocking) of main countries and cities have had to face, presents them with the challenge of evolving from provider of broadband and mobile connectivity to provider of IoT services—something that should came as seamlessly integrated in the environment as possible.

These IoT services will initially be aimed at some form of digital surveillance and diagnostic—for health and virus transmission chain tracing and for the overall flow and space management continuous monitoring. Telco could however leverage these "quick win" offerings. For example, they could start offering AI solutions able to optimize solutions in merchandising, following the example of Vue.ai, a unified visual AI platform aimed at supporting the retailing industry, using image recognition and data science to support merchants in their product proposition, catalog management and customer intelligence, and overall journey personalization, as shown in Table 6.3.

In so doing, they would start positioning as offline data aggregators and to develop into an advertising publishing platform, able to support merchants and corporates in reactivating their demand. Data are captured in shops, or in the street or other public spaces and mined to analyze behavior and emerging trends, with

Table 6.3 Examples of artificial intelligence for merchandising

	Rationale	Opportunities
Vue.ai	End-to-end retail automation platform applying Artificial Intelligence to retail. By using Image Recognition and Data Science it is possible to extract catalog data, analyze it with user behavior and help your marketing, product and cataloging teams get actionable insights that improve customer experiences, drive conversions and reduce costs	**Catalog Management** Extracts product data using image recognition and attributes products in detail. Merges inventory data with user behavior to plan, manage and forecast inventory **Visual Merchandising** Feeds data across channels, to showcase the right product, with the right styling at the right place, to the right cohort of users **Customer Journey Personalization** Integrates user behavior data to product data and personalizes across channels and feeds all the information to all decision makers

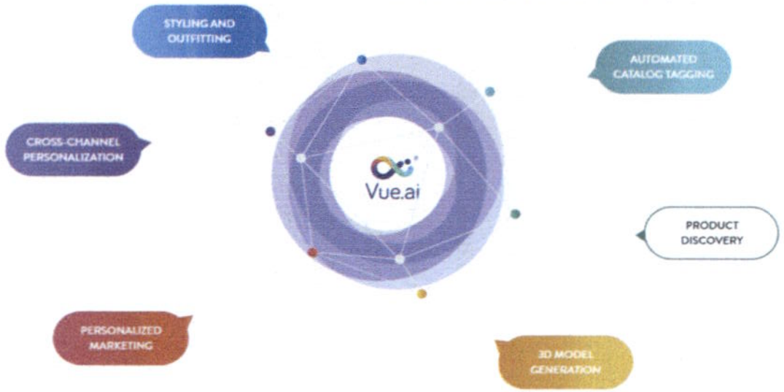

Source: Vue.ai

AlixPartners

targeted advertising delivered to customers via screens, dynamic banners, or their smartphone. The offline data gathered could then be segmented into meaningful clusters or anonymized at single client level to be compliant with the GDPR or explicitly ask for the sign-in of the customer (in exchange for a value-added service, a discount, or even some immediate cash back). Such data could become a valuable intangible asset to monetize, in compliance with GDPR, by generating fees as info-provider, based on clustered data and anonymized profiles of analysis. Also, an alternative business model to monetize this offline data would be for telecom players to support SME and small business as advertising, "cy-phy" channels, targeting end clients with tailored promotions. They could do it for a technology fee or as revenue share.

This disruption and the challenge to move from IoT service provider to offline (and online) data aggregator may however not be enough to reach a strong competitive positioning in the cy-phy setting of the "sustainable city" of tomorrow. In fact, not only the offering of mobile and broadband connectivity is potentially facing a destiny of competitive irrelevance (with provider of content having had already most the winning hands). But also, the positioning as IoT and related services provider is potentially under threat because of the commoditization of this hardware and the fierce competition on the IoT-based services offered by more agile start-ups. And even the positioning as offline data aggregator and monetization agent could not be that sustainable in a scenario characterized by the widespread adoption of "open data." Open data could in fact happen sooner than expected as the combined effect of two separate forces. On one side, the rule of law could force that approach, extending this from the European Union financial services industry to most geopolitical blocks and across sectors (namely, with big digital players included). On the other side, open data is now being independently promoted by market leaders such as Microsoft—with fintech already following. As data will progressively be made available and for free, at the whims of the original individual owner, most of the value that had migrated from connectivity to content and then back to IoT and network managers and big data aggregator could then progressively move towards the ones that are able to mine the (open) data available and turn it into "serviceable truth" to solve hard, relevant problems. Players like Microsoft will then not be doing this just for the sake of it but turning the idea into an opportunity to do more business. For example, Microsoft has teamed up with telecom providers like Verizon, Vodafone, and Deutsche Telekom[6] to help them roll out new dedicated 5G-based networks to business customers in markets such as manufacturing and logistics. As more open data is available and businesses look to 5G to potentially transform their operations, minimizing delays in transmitting data will become more and more critical, opening further opportunities for the players, like Azure, in the cloud business, that could establish and supply a network of data center that need to be placed physically near—hence in the main cities themselves.

[6]Will 5G era make it third-time lucky for Microsoft in telecoms?, Nic Fildes, *Financial Times*, November 16, 2020. Microsoft's 5G telecoms come back, Chris Nuttall, November 17, 2020

6.9 At Your (Digital) Health

Wealth optimization opportunities for cities (and then nations) are one of the main goals we started from, to drive our discussions and analysis on the potential roles for finance, real estate, and technology. We then discussed how, through a more radical redesign and transformation encompassing these three dimensions and more, such a wealth could be made even "smarter," meta, and hence more sustainable. Financial wealth is not, however, the only critical goal to watch as well-being, and, more broadly, psychological-physical wellness, is of essence. Wealth-being is then, we wrote, the summary of both, with interdependencies that are not just synergic, but truly critical, as wealth without health is of little comfort and vice versa. Also, we could argue no financial wealth at systemic level is sustainable if well-being is not shared by most people and vice versa. Wellness can mean many things, from mere survival to getting over a critical or less critical illness to feeling really good and great, and it should include both physical and psychical dimensions (not many of us have felt good during the "lockdown," even if physically protected by the pandemics and from many other minor illnesses—overall, the rate of non-COVID-19-related illness has also collapsed during the "lockdown"). Indeed, the opportunities for an improved and optimized well-being for individuals that could be associated with smart and meta cities are almost limitless. A "wealth-being" oriented city is first of all based on extended data capture that requires easy, cheap, and performant connectivity. In fact, connectivity provides the foundation of smart city services, and it can act as an enabler of smart healthcare and well-being as well. With such an infrastructure, communication between residents and healthcare and well-being professional counterparts and participants get easier and easier; hence the services towards residents are consumed in a more timely, personalized, and efficient way.

Personalization is, intuitively, of even greater value in these fields and even more so if associated with the psychological dimension of it, and this would reflect immediately on the financial wealth dimension as well. Who wouldn't be ready to pay more for a house that protects the health of our family (e.g., think of "senior living" neighborhoods for our parents) and our own (with real-time monitoring of basic functions, automatic sterilization and cleaning of rooms of viruses and small pollution particles, fine-tuning of temperature and humidity based on our state of mind, and so on, and from all this to illness prevention, first aid, support in recovery to a total "feel good" state of mind, the step is short). Ubiquitous connectivity, via integrated and seamless healthcare, embedded in the space and real estate we live in would allow to connect in real-time data collected from smart devices and sensors to extract valuable insights—not just on our physical state, but psychological as well—as analyzable via our tone of voice (as already done by Amazon's Alexa), the way we move our eyes or hands and eyes, psychometric AI is already able to read our face expression and infer our inner feelings and likely reactions to a given proposition. More basically, the technology that is already available can play a foremost role in real-time healthcare observation and help in early detection of health issues. Specifically, AI applications are already applied to perform tasks like analyzing laboratory tests, X-rays, and CT scans and perform data entry. AI-based apps are also in use to access the current medical condition of patients that can provide assistance in medical consultation, leveraging the knowledge accumulated via learning with experts. Finally, block chain technology is helping in redefining the methods of maintaining and populating electronic health records (EHR) as well and for linking them to other services like payments and insurance (upon the consent of the individual owner of the information). This may look already as big advancement vis-à-vis the scrambled, unreadable, paper-based diagnosis and receipt traditionally provided by family doctors.

Indeed, business-wise, we are still living in the Stone Age of digital healthcare, but with high expectation of some huge and short-term disruption to come, at scale. Big digital players and health tech start-ups are all planning to disrupt processes and practices, even if the "earthing" of those new practices clashes with the traditional legacy of the industry and with a number of regulatory constraints that are slowing down the pace of change that could already be allowed by the technology. The supply-side pent-up revolution, led by players like Google and Amazon, among others, is also increasingly matched by the pent-up demand of customers—all citizens potentially, as health and holistic fit are already recognized as part of the continuum of a holistic well-being proposition. We have observed how, in today's "smart" and hyperconnected cities, users expect an instantaneous and seamless flow of data. Many industries have then adopted, or are beginning to adopt, the required technologies to guarantee their users' expectations for instant information and (tomorrow) personalized solutions to the diagnosed issues they commonly face. As of today, the healthcare industry has lagged behind all this and for a number of reasons, including its high level of profitability and long investment and profitability cycle (for big pharma) and the inherent regulated nature of many parts of the industry—partly private but also largely state owned and state managed. Legacy systems, from an IT and data perspective, have also revealed very difficult to change and burdensome to tackle and even migrate onto more digitalized and "smart" platforms. Most processes are also slow, hospital-based, and made unnecessarily complicated by the bureaucracy and compliance implied and required by the role of the State (in Europe) and by the risk of being sued by patients for health damages and huge financial settlements (in the United States). Health data contained in legacy systems is also typically siloed and difficult to share with others because of varying formats and standards—with data landscape too fragmented and ill-suited to fit well with the instantaneous needs of modern users. As a result of this, stakeholders are incentivized to keep their own records, and no single version of the health-truth of an individual exists.

6.10 Poor Patient

Eventually, in most health-related organizations, starting from hospitals, the patient are as far from a client-centric approach as it could be, with healthcare often looking more as a set of products than an holistic service built around a person and with disjoint physical and psychical dimensions. The lack of a patient centric approach may be a heritage of a paternalistic relationship between healthcare professionals and patients. Instead, with changes in culture and the availability of digital technology, it is now considered reasonable to seek a second opinion, and patients are expected to contribute to decisions made about their treatment choices. In most jurisdictions, patients have the right to choose where and when (and increasingly, how) they receive their care and related data. Thus, with patient mobility comes the need for information mobility. In order to be provided with the best care, patients not only can but also must have the right to control their own data, for confidentiality reasons but also to reduce the incidence of medical errors of a procedural nature. For example, according to the *Financial Times* (Neville 2020), European data consistently show that medical errors and healthcare-related adverse events have occurred in 8% to 12% of hospitalizations. While 23% of European Union citizens claim to have been directly affected by medical error, 18% claim to have experienced a serious medical error in a hospital and 11% to have been prescribed the wrong medication. Evidence on medical errors shows that 50% to 70% of such harm can be prevented through comprehensive systematic approaches to patient safety, mostly based on, now largely available, digital records and administration and digital-based, AI-enhanced diagnostics. Also, telemedicine was already a clear trend in a before COVID-19 world and will just accelerate in an after COVID-19 one, with several applications already available, as shown in Tables 6.4 and 6.5.

Not only that. Apart from avoiding procedural errors or others driven by the inherent fallibility and unavoidable subjective bias of human experts, the opportunity for healthcare coming from data and AI can be related to the R&D and innovation in the pharmaceutical industry. Billions are now being invested into mining patient records in a bid to aid drug discovery (Neville 2020) and accelerate, in the case of COVID-19, the testing and regulatory approval of a vaccine. In fact, big data and AI could be the missing link that could allow for quicker and more effective R&D and experimentation for approval to then unlock more effective cures, bigger markets, and higher profits. In this big data and AI competitive game that could have such a great role to ensure the sustainability of cities, from both an health and a wellness perspective, big pharma and big digital will almost inevitably end up colliding, maybe joining forces along the way as they try to get access to clinical records to mine insights, through the analysis of vast troves of anonymized patient records. The payoff for society would be huge, as this competitive advancement would allow to expand the number of patients who can benefit from existing medicines, or even to unearth entirely new drugs, more quickly and on a more economic basis. In fact, big data and AI would allow to improve the pharmaceutical industry's poor record on productivity, accelerating drug discovery and regulatory

Table 6.4 Remote healthcare systems expected to experience high growth in future years and with COVID-19 accelerating this

Market

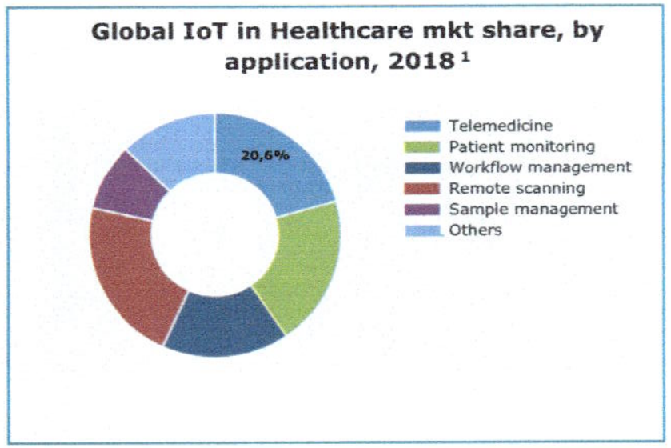

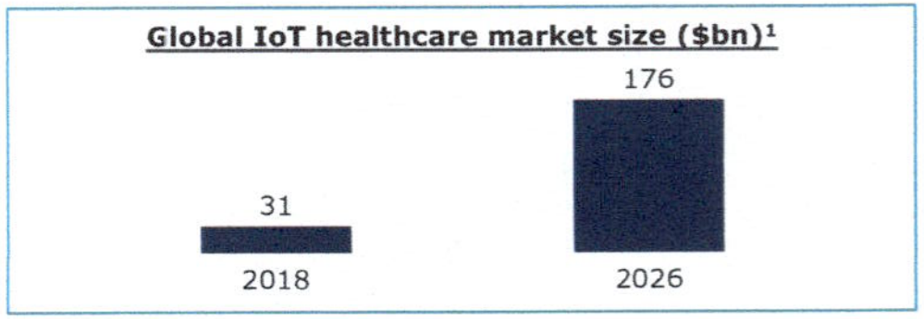

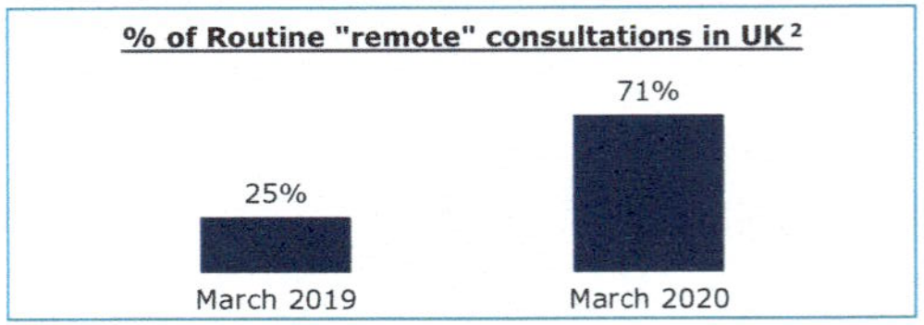

	Remote consultations	Chronic conditions / discharged patients
Opportunities	Opportunity to provide remote consultations through video systems and phone with following benefits: • Reduced costs • Simplify patient communication • Reduced time	Opportunity to monitor remotely chronic conditions with data transmission from patients to medical care providers resulting in: • Speeding up key decisions • Accurate and constant monitor of health indicators • Possibility to tackle health problems early

Source: [1]Fortune Business Insights; [2]Financial Times

AlixPartners

Table 6.5 Different mobile apps are currently being developed to support remote medicine—market examples

accuRx. Video consultations Patient questionnaires SMS a patient 2-way patient conversations Send documents via SMS	e consult Online consultations Patient questionnaires NHS self-help information Symptoms check
babylon Video consultations Patient questionnaires Healthcheck Health monitoring	KRY Doctor online appointments Open 24/7 Adult and child appointments Registered doctors and psychologists
DOC 24 Il medico sempre con te Video consultations Patient questionnaires Healthcheck Health monitoring Medical alert	tpp NHS login Video Consultations Appointment Management Medication Ordering Direct communication

Source: company websites

AlixPartners

approval and reducing research costs and failure rates. Two major shifts are potentially supporting the adoption of this big data/AI approach and with a typically cyber-physical approach, in the context of a "smart city" with extended IoT networks, performant connectivity, and twin replicas of individuals and their cohorts.

On one side, the demand from insurers and other health-related payers of last resort (including the State) is getting stronger and more sophisticated. New tasks are also including a more detailed and objective evidence on the relative performance of a drug (given its costs) once it has been adopted for regular use. On the other side, the increased complexity (especially in cancer cases) of drug discovery is just requiring a larger amount of information and greater computational power, as a better and more tailored understanding of individual biology is required to drive the search for more personalized treatments. Novartis and Microsoft have, for example, joined forces to develop drugs using AI, and an estimated 25% of all AI funding globally is related to health and pharma, with incumbent also investing in new health tech start-ups (e.g., Merck is investing 40 Mn USD in TriNetX—a clinical data and analytics company, with Roivant Sciences committing also the same sum to Datavant—it connects fragmented heath data sets to aid R&D). The overall, main idea is about augmenting and complementing clinical trials in areas where traditional data sources are poor or not enough and also, to do this in real life and not in a laboratory, with huge flows or real data captured in real time while observing real behaviors and in the context of real-life situations, from weather to food, to working or leisure-related activities. Novartis, which is defining itself as a data-science company, is also increasingly looking to integrate anonymized data from clinical trials with other information about patients, such as their genetic make-up, to find "super-responders" that are most likely to benefit from new treatments. This would accelerate their go to market and reduce upfront investments in R&D and also support continuous monitoring as treatments are applied.

Other players from other industries are also looking into the opportunity. For example, international real estate developers and global investors are looking to run similar experiments and testing jointly with big pharma players and in the context of "smart city" developments—one of such areas could be Arexpo, at the outskirt of Milan and designed as innovation center of excellence and built around the interaction and fusion of physical and digital spaces. Other players from the telecommunication industries are also considering to play a role, including Orange in France and Telefonica in Spain (Table 6.6).

6.11 Googling "Health"

Alphabet, owner of Google and one of the most famous, wealthy, and encompassing global big digitals (with interests spanning from its search engine and online advertising to e-commerce to AI and cloud computing and even to "smart" real estate developments), is, not unexpectedly, also looking to enter and revolutionize the healthcare and well-being industries, leveraging its unique abilities to capture

Table 6.6 Telecom operators have started to enter in the health space with different approaches—examples

Orange - France	Telenor - Bangladesh
Orange offers platforms to send mass texts and emails to communicate with health personnel in hospital. Solutions such as secure ID management enable caregivers to access hospital software to perform their duties. Hospitals can also share the number of available intensive care beds with regional health authorities. Connected medical devices are then able to monitor patients.	Telenor **launched Tonic**, a service that provides different digital health services from health information to a modern health line and medically-related financial benefits. Tonic was launched first in Bangladesh via Telenor owned Grameenphone, the country's largest operator with over 57 million customers.

Telefonica - Spain	Telstra (Telstra Health) - Australia
Telehealth: access medical agenda and send medical parameters and symptoms **Telecare:** this service gives the possibility of being attended to and located with a wide range of mobile devices featured with mobile teleassistance **Health IT:** facilitates collaboration among professionals in the diagnosis and enables retrieval of patient information by doctors remotely **My Health:** possibility to receive help in emergency situations and contact over the telephone for non-urgent issues	Connect health information, clinicians and consumers. Services offered are: • Phone & video consultation • Hospital telehealth • Patient in-home triage management • Home telehealth 

Source: company websites

AlixPartners

and analyze data, to find relevant information and intelligence for quick monetization. Indeed, its stated aims are higher as Alphabet is also looking to, literally, 1 day, save our lives (Kuchler 2020b). According to official statements, its aims are high and laudable, as it plans to fight medical disinformation in search results, create tools that could be used by thousands of doctors, and improve the accuracy of diagnostics with technologies like computer vision to read X-rays and the like. Of course, all these new technologies would also be used to disrupt and transform not just the offering of the health industry (a $ 8.7 Trn global industry, and growing), but also its sustainable business and profitability models and with huge opportunities for big digital players like Google and other, smaller but more focused health tech players. For future winners, it will become critical to leverage both physical and digital means and offline and online "value chains" (eventually, most of the health examination, diagnostics, and solution services could be done online, but not a surgery, a reanimation, or manipulation of a broken leg). All these should leverage the infrastructure of cities and hospitals based in urban and suburban areas (for the offline part) according to a more "holistic approach" to health and well-being—supporting us anywhere, anytime, and almost in real time.

For traditional players and, even more importantly, for new entrants, a key issue to address would be trust. Citizens would need to trust traditional players to let them capture their offline, functional data and during the normal course of the day and let them use their online data. Should these players be big digital companies, they would have already a huge amount of data to match—to analyze their lifestyle, diet, sleep behavior, etc.—so much important given the increasing relevance of the psychological health of hyperconnected people with all their connected disturbances. They would need to trust "big digital" companies or unknown "Heath-Tech" ones even more. In fact, trust is, for companies like Google, one of the main challenges to face. In truth, "big digitals" are scoring among the lowest in the ranking reporting Americans' willingness to share health information (Kuchler 2020a) (11% on average, but with Google as the company they trusted most among them). For them to really take a leap forward in eHealth, patients (citizens already classified as such) would need to trust "big digitals" to let them access their medical records, mostly still paper based and far from being digitized, let alone organized according to a "single point of truth" (a single and unique repository of all our examinations, prescriptions, medical opinions, etc.). Google is trying just that, treating patients (as in the case of its partnership with Ascension Hospital) to let them access their sensitive data. The "single point of truth" repository for digital health information would organize all the relevant, old data, but could also be developed with capabilities to mine them and to keep feeding new ones into its AI applications, via link with wearables and city sensors, and quantum computing engines that are able to analyze such a massive amount of data in real time in a 5G environment.

Such a unique eHealth repository could then be able to couple hospital records with all kind of IoT, including smartphones, genomes, sensors such as glucose monitors, and even apps that record eating habits (e.g., with an IoT inserted in your intelligent teeth or glasses). Digital watches are already able to detect and map your heart rate frequency and behavior, the level of oxygen in your blood (from your pulse), and even your perceived happiness or depression (looking at your face by video). Google has big ambitions on this kind of opportunities and has recently bought Fitbit that produces bracelets that monitor heart rates and other health parameters and with huge amounts of data with it. It is also leveraging Google Brain and DeepMind for their AI applications and Google Nest, to track sleep and lung functions to monitor and understand our wellness. It also owns (as Alphabet, the parent company) Verily (a life science division) and Calico (a unit researching ways to slow aging). Even more important is its development of innovative ambient sensors, including bedside devices, under-mattress sensors, and others integrated into toilet seats able to allow fast and seamless examinations that would then get mixed with the Fitbit ones (already able to measure our heart rate and blood pressure) and with new noninvasive health monitoring devices that will come to market as wearables, from contact lens to intelligent teeth and so on. On top of this, Google could also use unconventional sources of data to understand health as behavior and lifestyle driven, combining medical records and IoT fed information with web-based social data describing our behavior online (and, indirectly, offline) and analyzed also through our Google searches—if we are spotted "googling" health too many times, that would already be a meaningful indicator.

6.12 Healthcare of One

Smart cities are opening up almost limitless possibilities in terms of eHealth that, if captured and realized in full, could greatly contribute to their extension and development into meta and then sustainable cities. It could even allow to develop a health proposition that puts the individual (and even before he gets into a pathological condition) at the center and with personalized solutions to optimize his health and wellness. Thanks to digital innovation and hyperconnected cities, medicine and healthcare propositions could also become something of one—tailored specifically at each of us and to address our own, unique needs, as we are connected in real time and 24/7 in the city. In turn, this healthcare of one would become a critical dimension for the design and development of sustainable cities and—potentially—a leading discriminator for their competitiveness for businesses and attractiveness for citizens as allowed by big data, IoT, and AI.[7] This would also support the innovation and development of medicine itself, as continuous, hyper-granular, relevant flows of data would become available to allow for this to become more personalized and precise. Achievements in the study of human genomes are certainly contributing to such development, but genomics is not the only source of new personal health-related data. Just as genomes are unique, so are the lives that all genomes-carriers lead as they vary and are observable as take multiple forms in the city, individually and as part of homogeneous cohorts. Offline data on actual behaviors of people working

[7]Populations of one, *The Economist*, March 14, 2020

and living in the city, captured in real time and digitized for storage and mining by AI super apps, would help a lot in providing further truth: we are what we are because of our genes, but also because of what we do and what we want to achieve, given the business and societal contexts in which we live.

This information and intelligence enrichment coming from the IoT networks of cities, their fast connectivity, and AI super apps does not just cover observable behaviors and basic functional parameters, such as the blood pressure, heart rate, or level of glucose, that can be monitored by wearables. It can come as well in other forms of data about individuals, regarding, for example, their molecular information from medical tests, electronics health records, or digital data recorded by cheap, ubiquitous sensors (as the ones in the toilette example), making all things much easier. Even before the widespread adoption of wearables and "inside out" sensors, which could be installed in our homes and toilettes, the first opportunity to gather and analyze new troves of data are coming from ubiquitous smartphones. Almost 4 Bn people carry smartphones that can monitor physical activity, and, by 2022, over 1 Bn people will wear watches or bracelets that can monitor their heart rate. Much more is to come through the development of "smart," more digitally connected cities.

An obvious health service area that should be on the rise and experiment sooner with the new technologies available is telemedicine, something that was already on the rise around the globe, well before the pandemics, as societies struggle to deal with aging populations and the rising chronic disease burden that comes with it. Digital health solutions such as telemedicine are then becoming critical for driving efficiency and reducing costs—and could get a boost from the development of hyperconnected smart cities, leveraging IoT, big data platforms, and central, or distributed, AI analytics. In fact, in order to succeed, systems, devices, and data need to become seamlessly integrated. Payoffs are potentially huge, with the global telemedicine market valued at USD 40.11 billion in 2018 and expected to reach USD 148.32 billion by 2025 with a CAGR of 20.54% over the forecasting period, according to *The Economist* (and based on a forecast developed before the pandemics). Telemedicine and the healthcare of one could then become a single, unique, and historical challenge, with multiple changes required on both its supply and demand side and more likely to be successful if undertaken in the context of a "smart city." On the "supply side," new digital technologies (including in these wearables and nanotechnology, sensors, quantum computing, and advanced AI analytics) could greatly increase the efficiency and effectiveness of healthcare services. From software solutions that could be used to manage consultation bookings or bed management in hospitals, to complex AI-models to help radiologists detect cancers from X-rays (being tested by Google), to robots helping surgeons during surgeries, to biochemistry pilots aiming at finding better treatments for patients, the potential field of groundbreaking use cases is vast.

Also, on the demand side, developing a healthcare approach that is truly "patient-centric" in its approach, allowing individuals to get access to services in a seamless, more secure, personalized, and efficient way, should also become a critical target that could be more easily achieved leveraging digital tools and in the context of a smart city. From eHealth wallet solutions that improve data privacy to IoT devices that constantly monitor and advise patients to adopt better habits or call first aid hotlines, in case of a body-failure detection, everything could improve materially the patients' lifestyle. The storage of health-related information would also allow the accumulation of knowledge over the entire connected ecosystem—from the market participants sitting on both the supply and demand sides. This would then support the testing and application of the most suitable business models over the various interactions happening in the ecosystem. New business models could include, for example, personalized pricings when underwriting insurance policies, patients' data offered for sale to the pharma industry, subscription models for software and hardware healthcare tools, personalized medical practices, and so on. Some of these business models could even come as "embedded" in the smart city, neighborhood or even at building level—when, for example, marketing new real estate developments aimed for "senior living," all coming with a new plethora of smart devices that would help monitor, alert, track, and support the growing senior community, whether they're living in smart shared buildings or in their own family homes. They could track vital signs, sending emergency emails or texts in real time to care providers if something is offtrack. They could detect body facts and values and proactively suggests personalized diets and workout plans, reducing consultations (or moving them fully online) with family doctors and nutritionists. They could also raise warnings on things like a low level of movement and abnormal sleeping habits, allowing care teams to remotely track behavioral patterns and check on patients (third age but not only) if needed. Data from wearables could then be sent to the cloud and then analyzed and measured to find trends and insights for doctors and family members (even for the prevention of the latter). Finally, pacemakers and glucometers could be even implanted directly into the senior's body to track vitals, ensuring that the technology is always accurate with periodic testing taken offline.

6.13 An eHealth Wallet for Sustainable Cities

A critical step to make this vision come true will be represented by the ability for smart (and aspiring sustainable) cities to build eHealth digital wallets that become the single source of truth for our health conditions and—potentially—an open source of anonymized data that is available for free and supporting innovation and development in medicine, pharma, and healthcare. In fact, in the current healthcare system, patients have their health information spread over multiple systems, hospitals, networks, and potentially even countries. There are multiple fragmented records of the same patient, held at different institutions—all with their own snapshot of the patient's health during their interaction with them. As a unique point of reference for patients (who, as of today, have probably their health data scattered in multiple paper folders), it could grant access to hospital and doctors on the basis of a clear set of rules. It would then ensure an effective proxy for the single source of truth that could benefit all market participants, with a number or easy, effective, and intuitive use cases to deploy (e.g., avoiding repetition of exams for second opinions, etc.). The platform (or eHealth wallet, or folder) would order and filter these records into a chronological order and the specific categories above to aid data handling. Such categorization would make the records more accessible and understandable for patients and facilitate researchers as well, in seeking out the information that are more important to them. The user will then be able to transport their records as they travel. Healthcare services would become borderless as telemedicine services allow users to interact with clinicians in other countries and become available to multiple counterparts.

Such a platform could then develop as integrated for the whole ecosystem—getting in touch with existing health- and fitness-related applications and IoT devices (e.g., Fitbit, Apple Health, carbon and monoxide IoT detectors, etc.) and open to third parties to build applications that can complement and improve the user experience of healthcare and medicine. Ultimately, the citizens-patients could be able to leverage their medical data to power a plethora of applications and services, and all interactions could be recorded, audited, and secured on a distributed ledger, supervised by smart contracts. The combination of a centralized platform with a distributed tracker would then enable the secure, fast, and transparent exchange and usage of medical data, aggregating accounts for a solid base of intelligence. The platform could then offer an API that would allow third parties to obtain and interact with EHRs (electronic health records), upon the user's permission. The advantages for all participants, in the setup of a sustainable city that can also master an eHealth wallet, would be obvious. While in present-day healthcare, there is a lack of transparency between all parties involved, with patients having no immediate access to health records written by medical professionals; with medical professionals that are only able to share data quickly within their own organization; and with insurance companies that are kept in the dark, unless they request patient data which is required for a claim. And, finally, with researchers that are forced to seek anonymized data from multiple intermediaries, which is both costly and timely. The eHealth platform would instead allow clinicians to communicate with each other with ease. Any clinician with authorized access to a patient's record would see its updates in real time. And any doctor with a browser and an Internet connection will then be able to access the users' documents shared with them. Also, the centralization of the records and the immutability of the distributed ledger tracker would allow insurance companies to make a more accurate assessment of an individual's health, and health premiums should reflect this. And researches that have been granted access could perform their data science over the data collected. All this would of course pose not just new opportunities but also great risks, in the form mainly of security—mostly cyber-related. As cybercrime around the world is on the rise, healthcare systems would be no exception as shown by recent high-profile ransomware hacking and with even more critical implications. In fact, the healthcare industry has more data breaches than any other sector and medical records are being stolen and passed on. Creating cyber trust on its eHealth digital wallet (or folder) should then become a priority for cities.

6.14 Strategic Forays in AI-Based Use Cases

Our quick overview on the potential disruptions that could happen across an initial set of industries, starting from financial services, TMT, and health could actually extend to all of them, including retailing and entertainment, transportation and manufacturing, and so on. In order to provide some potential exemplification on few of the business designs of the future, we have provided here a limited set of potential use cases where AI, and the monetization of physical and digital data, could be put at use to derive some form of competitive advantage. A first use case, in Table 6.7, is addressing the potential analysis of how companies' behavior is impacting climate change and how, based on their monitored behaviors, we could forecast changes in CO_2 and hence climate-related risks and predict its impacts. Based on this, a more granular set of policies (including taxes, but also subsidies) could try to change and redirect their behaviors on the basis of an optimal trade-off between costs incurred and expected benefits. Banks, telecoms, and real estate players would be expected to contribute a set of relevant data, but also big digital, online only players. Eventually, on the basis of a very thorough analysis of behavior-driven climate change, financial guarantees and derivatives could also be set up as tailored OTC (over the counter) contracts, with a view to align incentives and behaviors. Standardized, traded derivatives, supported by broker-dealers as well could also be set up to mimic the effects of a carbon tax on CO_2 emission or similar.

A second use case, also in the area of risk management and described in Table 6.8, would support the buildup of a more resilient and effective ecosystem for banks, finance, and real estate by leveraging on- and offline data. This big data, analyzed with the help of AI applications, would sustain the development of better intelligence on real estate value—not just examined on a mark to market basis, but also assessed with regard to its future trajectories and its main drivers of accretion, given observable behaviors and emerging trends "mined" from information. Such a real estate intelligence engine would help banks and other investors in their underwriting decisions, but also real estate developers in designing and executing regenerations or developments that would be more attractive to perspective end users, hence more valuable and resilient—given potential changes in the overall ecosystem and in the related supply and demand of real estate built assets. Other use cases in finance and related to the consumers would also be possible, linking banking and online advertising, lending, and e-commerce. Some of them are actually already being pursued by big digital players. For example, Goggle (through Google Pay (Lex 2020)) is looking to link the consumers' personal finance information with its global search and advertising businesses. It will let retail clients transfer funds, manage their budgets, and even open a bank account with a set of participating banks. Likely (and notwithstanding the antitrust lawsuit it now needs to face off with the Justice Department), it sees payments as a way to embed itself even more deeply in users' lives—it even includes new functions that allow the app to trawl through the clients' Gmail inbox and Google photos to look for receipts—a practice that some may judge as invasive rather than useful. Also Amazon is following suit, with a

Table 6.7 Evaluating impacts of companies on environment and assessing how climate change affects companies' performances

CHALLENGE	• Unprecedented **scale and scope of climate change challenge** and complexities tied to its **measuring**, due to: – **Limited usefulness of historical data**, typically of a structured nature – The **unreliability of current statistical, conjoint analysis-**based risk models • This is limiting companies' ability to **allocate investments to the right projects** in terms of CO_2 emissions and to **react** – and plan ahead – **to immediate risk threats** • **Financial institutions** have a large **role to play** and are increasingly called by policy makers, regulators and institutional investors to **act consistently**
ROLE OF DATA AND AI/ML	• **More reliable indicators** on who/what is causing CO_2 emissions and **more robust risk models** could be derived through: – **Collecting all available data**, but mostly unstructured / social-behavioral ones, available in real time through satellites, web and an extensive network of IoT – **Using AI/ML applied analytics** to master the high number of variables and data to make the absence of historical data less relevant
STAKEHOLDERS INVOLVED	• **Owners of IoT and telecom networks** across Italy • **Real estate developers**, in charge of large **new development** projects in urban areas • **Banks** as main lenders of a large number of **new development projects** • **High tech player** as provider of computing tech and applied analytics
PROPOSED SOLUTION	• **Framework to assess how companies and banks specifically are supporting the transition to a low-carbon emission economy**, considering a set of criteria from passive (e.g. sustainability goals and transparency), to active (e.g. sustainable financing strategies and products) and superior (e.g. zero financing of emitting industries in the balance sheet) • Each criterion can be used to develop **internal new scoring models** as well as **communication strategies** for markets and extended stakeholders

AlixPartners

Table 6.8 Building a digital ecosystem for banking, finance, and real estate and to support better credit scoring and selection of loans

CHALLENGE	• Banks tend to act as '**hedge funds**' with a long position on the Real Estate cycle, without really understanding the inner dynamics and implied volatility (and risks) at single asset level • Real Estate is **one of the least understood and analyzed investment asset class**: the price of a RE asset is driven by **location** and this one by its relative **attractiveness**, which cannot be measured only using traditional and structured data (e.g. financial statements, market reports, POS cash flow and credit card spend)
ROLE OF DATA AND AI/ML	• The opportunity is to **capture all kind of 'unstructured' big** (or smart) **data** to develop a complementary and more forward-looking view on the value of the Real Estate • Examples include **social ratings** and feedbacks on corporate clients provided by customers through social platforms, **micro-areas characteristics** and **in-store measurement** of customers' activity • **AI/ML tools** would **analyze the data and derive intelligence**, e.g. industry specific business performance metrics, micro-segmentation through enhanced customer demographics / behavioral profile, analysis of competition density
STAKEHOLDERS INVOLVED	• The **bank** could act as the **main actor** in the **definition of an ecosystem vision**, aggregating stakeholders (e.g. final clients, RE developers, RE agencies) needs and optimizing real estate wealth-being
PROPOSED SOLUTION	• To transition from 'hedge-fund like' to asset manager, broker – dealer and then wealth manager, banks could **start from three use cases**: - Integrated credit lending dashboard that brings together **different sources** of data to support the bank throughout the various phases of the **credit value chain** - Mortgage match-making solution that analyses **structured and unstructured data** to better understand the **"quality" of the area** and the forecasted **"solvency" of the borrower** - **Open Data for RE and services for residents, leveraging domotics** and **smart home tools** to obtain **residents data** and be able to serve their needs more efficiently

AlixPartners

tie-up with Barclays' credit card business in Germany, aimed at offering seamless customized shopping and payments services. Essentially, through this partnership, that Jes Staley, CEO of Barclays (Tett 2020), is considering as one of the most important things for his bank in the last 5 years, Amazon and Barclays are looking to link data with AI analysis to underwrite credit and predict what customized services clients will want next. In theory, and even more within the context of a sustainable city, it could be beneficial as a way to "democratize finance," as Mark Carney, former governor of the Bank of England, has observed. It should allow, in the logic of the synapsis bank, to offer consumers more choice, better-targeted services, and keener pricing.

Use cases leveraging on offline data could also be designed to support the creation of new revenue sources and not just for risk management purposes. The following use case (Table 6.9), as an example, is potentially applying the highly successful online-only digital advertising model of players like Google and Facebook to the offline world as well. The idea is simple. As online customers' funnels can be analyzed, mapped, and leveraged to provide a targeted advertising of one (meaning, we receive the most targeted and spot-on advertising, in the right moment and based on our inner needs and pent-up demand, as "discovered" through online data analysis by AI apps), the very same could happen in the offline world, if well-designed IoT networks are implemented and managed at the physical point of sales of merchants. Merchants could capture data of offline customers as they behave and make consumers decisions in the atom world; mix them with online, bite-based information; and derive a superior intelligence on who they are and would like to be, as they consume their way out towards their aspirational self. This could help as customers are monitored as they walk in public areas, enter into shopping centers and entertainment venues, and move around, attracted by a specific lay out, product, sound, or even smell. As they look around to discover the physical place they are living in, they give away themselves, opening the doors to targeted, real-time advertising (on top of customers' search and data science analysis). We are hot and sweating? Maybe an advertising should invite us for a drink at the nearby coffee shop, with a special discount via a digital coupon that has already been received by our smartphone—but expiring in 5 min. We could also link our patterns of behavior to more structured and longer-term loyalty and cash back programs and receive a consumer finance proposition that is based on a credit risk scoring and risk-adjusted pricing also based on our analyzed, observable behavior, as we live in the city.

This customers' funnel analysis that is applicable at point of sales, shopping centers, and leveraging the digital payments and cash register technology of the merchant could also be replicable in an entertainment, open space, public setting, where large crowd of people get together and where it would become possible to study not just the behavior of the single consumer entering a shop to buy something, but also heterogeneous cohorts—lots of apparently uncorrelated individuals that however get together to live with great joy or moment of sadness in the space of a few minutes. Think, for example, of soccer stadiums that are crowding together over 50,000 people from many places and different culture, profession, and lifestyles—but all sharing a common passion for their club of choice and behaving like fans as a

Table 6.9 Designing and addressing the opportunity of a "cyber-physical" advertising model for SME and merchants specifically

CHALLENGE	• **Great discontinuities** ahead: – Rise of "**smart cities**" and transition from models based on utilities (mobility, security, civic services) and big data platforms to **models based on ubiquitous diffusion and real-time connectivity** and technological solutions ("**meta-city**") – Offering centered around individuals and their **wealth-well being** – **IoT** with almost **unlimited charging** and **wearables** implantable in buildings and in human body to enhance physical and psychic abilities (e.g. contact lenses with info screen) – **Explosion of intelligence** derived from the use of aggregated data starting from the physical environment • The **winners** will be those companies that can **capture and translate online and – increasingly – offline** data to carry out behavioral analyses and commercial funnels • **Advertising and information provider** represent the **main monetization opportunities**
ROLE OF DATA AND AI/ML	• **AI** and **advanced analytics** as the **critical elements** to be able to **translate data** captured from IoT, web, phone lines and other unstructured sources in **comprehension of human behaviors** (consumes, socializing, productivity, health etc.) and its **valorization**
STAKEHOLDERS INVOLVED	• **Telecom providers**, as owners of **physical assets** and connectivity networks required for the initial phase of the "meta-city" development – and **data captured online** (e.g. phone bills) • **Payments providers** that have a presence at Point of Sales level for ensuring digital payments and capturing data related to the transactions being cleared
PROPOSED SOLUTION	• A solution could **combine "online"** and **"offline" assets to gather information** and use it in a **number of possible use cases**, e.g.: – **Cash registers and IoT offering** for merchants to analyze customer behavior and monetize it through cross-selling and enhanced marketing – **Merchants as sellers of offline marketing** – **Credit offered to merchants** from direct lenders based on their "enhanced" profile (e.g. data on offline store visits and customer behaviors)

AlixPartners

result—hence by definition in economic "irrational ways." A well-designed IoT network could capture data in real time and work out the feeling of the crowd, anticipating their needs and also the best moment to offer them something of value—based on their shared feelings and individual needs, as highlighted in Table 6.10. Offline and online commercial propositions could then follow up, proposing a further dimension to their cohort-based lifestyle, even cutting across people with hugely different available income and wealth.

A further and very different use case, as described in Table 6.11, could instead help us in our journey back to controlled normality (or to an "another normal") as the pandemic lockdowns are released and the virus is progressively constrained, monitored, and controlled, with the help of a real-time, IoT-based, digital monitoring—usually with the contribution of public, central, and local authorities and private companies. Anything from mobility in the city or across the country, the return to the office and schools, or our holidays and entertainment and use of the most current and routine (but severely interrupted) services would benefit from the continuous observation and analysis that is feasible only through the capturing of huge amount of offline data and its (also happening in real time) mining, to understand the trends and emerging risks and behaviors that need correcting. All this is of course posing serious questions regarding the ownership of the data, the protection of private and confidential information, and the correct, admitted end use, to avoid that a monitoring engine that is set up for health emergency reason is instead turned into a commercial money machine, with for-profit organizations investing on this just to gain an unfair competitive advantage or, even worst, into a tool for tyrants to master their servant citizens.

Finally, one last use case shown in Table 6.12 relates to the eHealth system that could be set up from the real-time aggregation of functional body parameters, from heartbeats to temperature to glucose levels, to support a tailored, just-in-case first aid proposition, but also a preventive diagnostic one, with well-being services also designed specifically for the person being monitored. This eHealth proposition could be location-independent (if IoT are used as wearables or similar) or more location-specific if some services are related to the place of living or working, potentially contributing to the attractiveness and value of the place (e.g., for senior living residential areas). This data gathering could also allow quicker testing of new pharma solutions, as people would be observed and monitored as they work and live and not constrained in a laboratory.

6.15 Serviceable Truth

The notion of serviceable truth, in the original conceptualization of Robert Nozick, is based on the very idea that the real-real, absolute truth is hardly attainable. Not only full objectivity if often impossible to reach, but even assuming that something objective-enough would be reachable and constant looks also as wishful thinking. Things (people, companies, institutions, and cities that host them) can hardly stand

Table 6.10 Designing a better user experience for soccer/sports fans and monetizing data by leveraging IoT, big data, and AI/ML

CHALLENGE	• **Soccer teams have a significant improvement potentials** across all dimensions, with low revenues across the match-day, commercial, sponsorship and other dimensions – this translates on a low brand equity value – Total revenues of Italian peers: Juventus 2.9x, Inter 2.1x, top Europeans at 7.3-6.7x • **Opportunity for soccer teams to design and build a better and more performing "cyber-physical" community**: – 2nd as social media fans among Italian soccer clubs – Fans joining the stadium steady/ marginally decreasing – Data, info-telligence and profiling potential largely un-tapped for cross sell and sponsorship • Soccer team **could better capture, integrate and optimize the fans' online and offline funnels**, optimizing cross sell and ADV online and offline, improving and extending the stadium experience for families in time and space
ROLE OF DATA AND AI/ML	• **Data capture and mining** can support the development of **profiled and integrated views of individuals** and **"cy-phy" tribes** (masses and cohorts), that swarm on-line and meet in the physical realm of the stadium
STAKEHOLDERS INVOLVED	• Potential franchises with **shops and hotels** around the stadium areas • Potential franchises with **media agencies** to further promote the soccer team and fans lifestyle • **Third party advertisers**, to which the ACM app could be opened • **IoT devices** producers
PROPOSED SOLUTION	• **Increasing convergence of cyber and physical** (cy-phy) dimensions is key **to design and build better and more valuable communities** in the lifestyle, affordable-luxury, entertainment and wellbeing ACM target market • Three potential level - **online-based; stadium-level** and as **extended real estate proposition** as franchising of shops and hotels and as part of resi-mixed used development and urban regeneration program

AlixPartners

Table 6.11 Managing the post COVID-19 "return to another normal" leveraging data, extended connectivity, and AI

CHALLENGE	• The **un-lockdown after the corona virus emergency peak** will be **gradual** and needs to be **managed carefully**: - **Back to work**, likely gradually phased-in and differentiated by specific risk categories hand with new ways of doing things that will be here to stay - **Mobility**, still partially limited and selectively regulated - **Social distancing** measures - Rise in **social unrest**, thefts, acts of violence and other forms of extreme protests • At the same time, there will be a **strong demand for services and products**, related to the **backlog of activities accumulated in the past weeks** and the willingness of the population to go back to the normal life, including all kind of non-necessary goods, non COVID 19 related surgical interventions / diagnostics, services to the person (e.g. hairdresser) and goods (e.g. plumbers) that require some physical interactions
ROLE OF DATA AND AI/ML	• **Existing data, connectivity services and artificial intelligence applications can be leveraged to manage the re-start,** e.g. to monitor individuals, places and activities, support Rule of Law if selective approach is pursued, develop apps as required
STAKEHOLDERS INVOLVED	• **Telecom providers** owner of geo localization data, towers, cabinets, data usage, technical vans • **Big digital,** owner of geolocation data, advertising investors, virtual classroom, cloud / computing • Other partners including **financials institutions, law firms, hospitals, healthcare companies, universities and education, entertainment clubs and gyms**
PROPOSED SOLUTION	• **Central authorities** could team up with selected partners to **leverage existing data, connectivity services and AI applications** to optimally manage the re-start • Several **use cases**, with an **ecosystem of partnerships** required for its implementation

AlixPartners

Table 6.12 Building a "smart" healthcare system in new real estate development to support urban regeneration programs

CHALLENGE	• In the current **healthcare system**, patients have their **health information spread over multiple systems**, hospitals, networks and potentially countries. There are **multiple fragmented records** of the same patient, held at different institutions with their own snapshot of the patient's health during their interaction with them • The **aging of the population** and the **growth in the residents' demand for smarter health and wellbeing** makes investments (or co-investments) in the digital healthcare an opportunity to generate new VAS driven revenues and increase and the attractiveness and value of the place
ROLE OF DATA AND AI/ML	• **Connectivity provides the foundation of smart city** services and it acts as an enabler of smart healthcare VAS as well • With such an infrastructure, **communication** between residents and other market participants gets **easier** hence the services towards residents are consumed in a **more timely, personalized and efficient** • A**n entirely new proposition of "smart health care" and wellbeing can be developed and executed**
STAKEHOLDERS INVOLVED	• **Real estate developers,** in charge of large new development projects in urban areas • **TMT players** as owner of geo localization data, towers, cabinets, data usage, technical vans • **Big pharma companies/ hospitals** as owner of R&D process and of health services/ solutions
PROPOSED SOLUTION	• The combination of involved stakeholders could **address**, in the context of the innovation district and new urban development, the **shortcomings and gaps of the current healthcare industry** • **Development of a «patient centric» approach, entailing holistic health care and wellbeing monitoring**, reporting and analysis, based on "always charged" IoT and with analytics correlating basic vital functions with lifestyle information, with health-to-wellbeing advisory and VAS offered on the basis of the intelligence generated by data captured and AI apps

AlixPartners

still, with their behaviors and related, produced data changing continuously and therefore subject to great variances. In fact, "invariances" are our second best options for using objective enough approaches to derive truths that are also serviceable enough—i.e., are kind of useful for the specific end use at hand that must be critical and relevant to justify a relevant value derived from this serviceable truth. For a telecom player (or actually, any kind of player) to become a winning cy-phy info-telligence solution provider, it then needs to define a clear set of high priorities and relevant and critical needs to solve for its target clients, potentially reinterpreting traditional categories based on this—even more so in a post COVID-19 world. This could allow telecom players to migrate value from other sectors (and vice versa). Take the transportation sector, for example, that is fulfilling our primary need of mobility—with trains competing with cars and both with flights and so on. In a cy-phy context, this need needs to be reinterpreted and extended given increased blurring boundaries. We can all get more mobile (given our ultimate needs associated with that) by better leveraging our cy-phy dimension, e.g., doing more 3D videocalls or travelling faster on bullet trains running in air-vacuumed tubs managed by AI algorithms or working while travelling with self-driving cars that are also guided by data provided in almost instant times via 5G.

Take also healthcare and well-being as two other large sectors that are ripe for disruption. In a pandemic-stricken, climate change-threatened world, healthcare—fulfilling our primary need of health and well-being—will need to be reinterpreted in a fairly extensive and cy-phy way. We can get more fit by training at home or when travelling (also, exercising as we travel, cycling across "smart" roads and bicycle lanes, or doing push-ups and trade-mills when on a videoconference call) and connecting with other people to enjoy the experience of training. On this, Peloton stand as a great success study on how a traditional training tool—the static bicycle—can be reinterpreted as a crowd aggregating hub, with hundreds or thousands of people joining in real time and across the globe the same spinning class, where cycling gets mixed with dancing and trainers are most often former disk-jockeys than professional cyclists. During the lockdown time, more professional techno-equipment providers, such as Garmin, have translated into the digital world classic races such as the Milano—San Remo or the Giro d'Italia, with the opportunity for amateurs to actually race (time clocked, power-measured and with a video simulation of the mountain pass they are climbing) against the pro, Vincenzo Nibali included. Pros themselves have engaged in digitalized, unofficial "monument classic races." Also, we can get health checks and real-time monitoring and diagnostic as well, and even proper healthcare when ill, limiting the time we spend in hospitals for testing or when actually hospitalized. The COVID-19 outbreak has also accelerated the development and adoption of similar digital apps. During the lockdown, more than 70% consultations in the United Kingdom have been remote, compared to 25% on the previous year (Venkataramakrishnan 2020). Digital triage is offered by companies like eConsult, which allows patients to send to their GP their symptoms in written form along with pictures. Or accuRx that offers video consultations and uploading of documents, text, and photos. And, as already discussed, all kind of telemedicine has also expanded its offer, to allow people to carry out testing remotely, with their GP's supervision via video link or with samples send by mail when more heavy lifting analysis is requiring—freeing the health (and transportation) systems as a consequence.

In short, no matter what the sector chosen as a starting point, most competitive offers could become blurred, with telecom competing with automotive, gyms, and healthcare providers, not to mention big data players, cloud specialist, and AI machine-based specialist providers. As the convergence of cyber and physical (cy-phy) dimensions progress to develop more proper and integrated views of individuals that swarm digitally and meet in the physical realm, cy-phy companies have a chance to play game-changing strategies—not only delivering the "Milano San Remo" race experience into your living room but also bringing to you (when actually cycling the Cipressa pass) the information flow that you could have when sitting in your office and the real-time medical testing that you can get by visiting a specialized medical team. Accordingly, winning companies could come from multiple sectors but will excel in managing these cyber-physical perspectives in the city and in identifying hard problems and pent-up demand to address and fulfill. Sustainable cities will then become the competitive battlefields for these cy-phy companies to understand customers' and communities' behaviors, build loyalty and brand equity value with respect to their ability to identify and solve relevant and critical use cases and map, and understand and control cy-phy solutions—which may be then designed and executed in partnership. As everything will change, banks, telecom, and real estate companies will all need to change—along many others, with their sector of origin becoming less relevant. Almost paradoxically, as cities will develop and become more sustainable, traditional categories to articulate the way they work will become less so.

References

Hannah Kuchler, Stanford medicine, Rock Health. Reported in Can we ever trust Google with our health data?, Financial Times, January 20, 2020a

Hannah Kuchler, Can we ever trust Google with our health data?, Financial Times, January 20, 2020b

Lex, Google Pay: checks and balances, Financial Times, November 19th, 2020

Sarah Neville, Why big pharma sees a remedy in data and AI, Financial Times, January 26th, 2020

Claudio Scardovi, Digital transformation in financial services, Springer 2018

Claudio Scardovi, Singularity bank, Bocconi University Press – EGEA, 2019

Gillian Tett, Artificial intelligence is reshaping finance, Financial Times, November 19th, 2020

Siddarth Venkataramakrishnan, Lockdown drives boom in healthcare apps, Financial Times, May 5th, 2020

Chapter 7
Investing in Real Estate: BC/AC

7.1 Real Estate BC/AC

Real estate, as the largest asset class globally, and the most visible proxy for the wealth of nations and of its main cities, is in a state of turmoil, with lots of uncertainty in world that will be marked by the most accelerated and profound recession in recent history—for most countries, the only comparable time dates back to the Great Depression of the 1930s and for others to the impact of the two world wars. Such is the water shedding impact of the pandemic that we could venture that for real estate—and for investing in it specifically—there will be a "before COVID" time and an "after COVID" one, with the all known unknowns that for sure will impact cities and their real estate and infrastructure with them. In order to understand what the future of real estate investing could be, in an "after COVID" world, it is useful to reconsider what was driving value and investing in a "before COVID" world, also on the basis of what we have been discussing so far on the future of cities. As a start, real estate investing is, as all other forms of investing, all about comparing a cash-out (a price, or the investment to be done on day one) with the net present value of all future cash-ins coming from the invested asset. Eventually, also a real estate asset can be evaluated on the basis of a discounted cash flow and taking into consideration a proper discount rate (associated with the level of riskiness of the specific investment, on top of other considerations regarding the prevailing risk-free rate of the economy and the risk premium the market would pay for macro asset classes, such as equities versus bonds, or real estate versus bonds). If also real estate can be associated with discounted cash flows, a good proxy for future payments is the attainable rent—hence rent over price (as a perpetuity) is a first index (or "rule of thumb") that could be used to estimate the value of the investment. If this "rent/price" ratio could be associated with the expected return offered from the supply side of real estate assets, two other common ratios could be effectively describing the willingness to buy on the demand side and their relative price inelasticity given

© The Author(s), under exclusive license to Springer Nature Switzerland AG 2021

C. Scardovi, *Sustainable Cities*, https://doi.org/10.1007/978-3-030-68438-9_7

limited supply, as shown (through regression analysis applied to real estate historical series) in Table 7.1.

The higher the available income for individuals, and the higher the lending supply (in terms of mortgages offered to retail and corporates), the higher the price the aggregated demand would be ready to pay for a square meter of some bricks and mortar asset. In fact, intuitively, we can either pay for a real estate asset through our own ability to save or through our ability to get loans—hence an "available income (and/or) lending supply over price" is also a good indicator for gauging whether the real estate sector is priced on a demand sustainability basis at around or near its fair value (as described by the future rents I could get if I rent the asset instead of just owning and using it). Regression analysis have proved that these ratios are pretty strong indicators of future price changes, with correlations of over 90% over time. Still, these ratios are not capturing the dimension of attractiveness of a place (why spend so much in Venice instead of a fraction in Dolo, just a few kilometers away?), nor the limited supply dimension (that is also driven by the exclusivity of the specific asset: if limited detached houses are available in the center of Venice, they will be worth even more). All these were true BC, but in AC world other potential impacts need to be considered. For a start, the rent of real estate assets could vary greatly, given the significant short-term or long-term shift in demand for offices and shopping malls, for example, vis-à-vis residential buildings or logistics. Also, the net available income could be significantly impacted for certain part of the population, given higher expected unemployment, serial bankruptcies in certain sectors, and so on. Lending supply could also be impaired in the mid-term, if banks are expected to face a new, long wave of NPL (non-performing loans) that will eat their capital and limit their mortgage offering as a consequence. All these factors are negative, but the limited stock of properties with certain characteristics (e.g., large residential flats with a garden or a terrace and temporary office facilities) could also work the other way around, as people seek to readjust to the new ways of living.

7.2 This Time Will Be Different

The history of macro and microeconomic, negative shocks and global crisis is usually played with the relevant role of five main culprits—starting, for example, with an asset bubble (typically, in real estate), a systemic bankruptcy (usually, in financial services), unsustainable expansionary fiscal policies (usually driven by election cycles), reckless monetary policies (also driven by political cycles), and, eventually, a powerful mix of greed, fraud, and sheer stupidity. Banks have mostly got burned by real estate through the last cycle and after the global financial crisis, but private equity investment have been proving attractive and resilient through cycles and with good performances over the last 12 years (and particularly so, in times of stress and high volatility, if the investment strategy pursued is of a more tactical and opportunistic nature, as shown in the analysis of Tables 7.2 and 7.3).

Table 7.1 Real estate valuation based on fundamental analysis

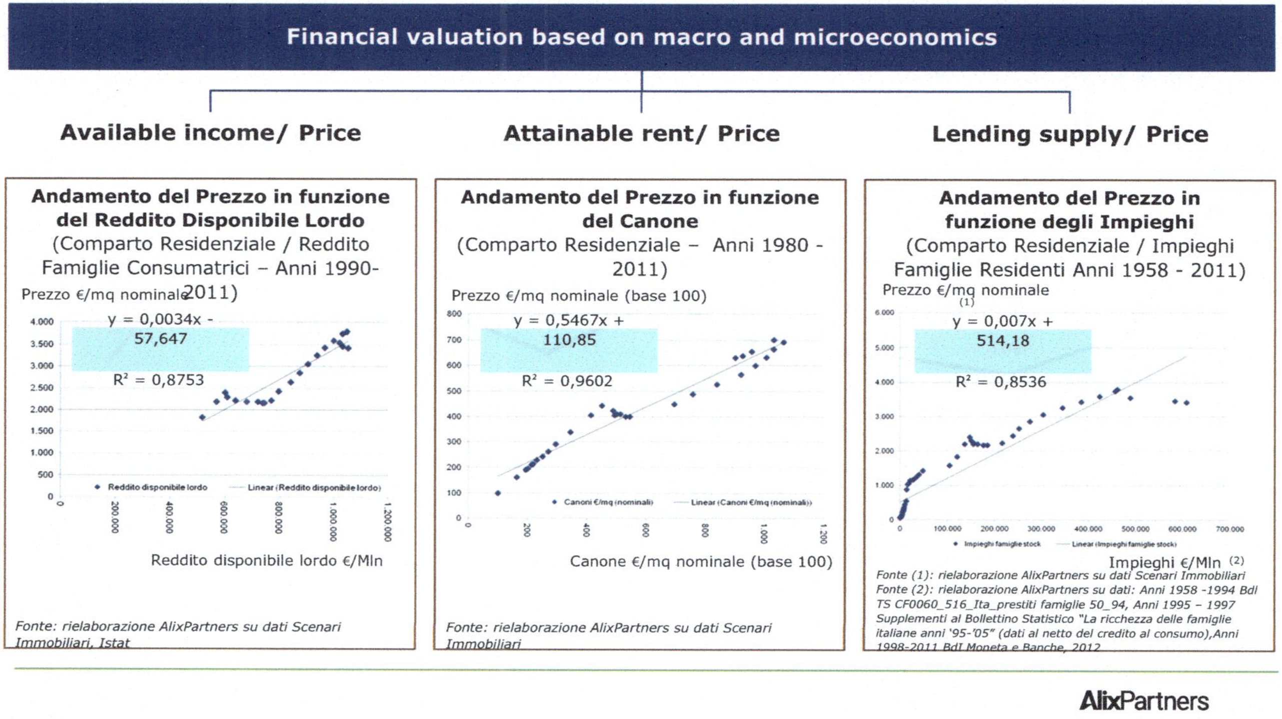

Table 7.2 PE investing in RE has been resilient through the cycle

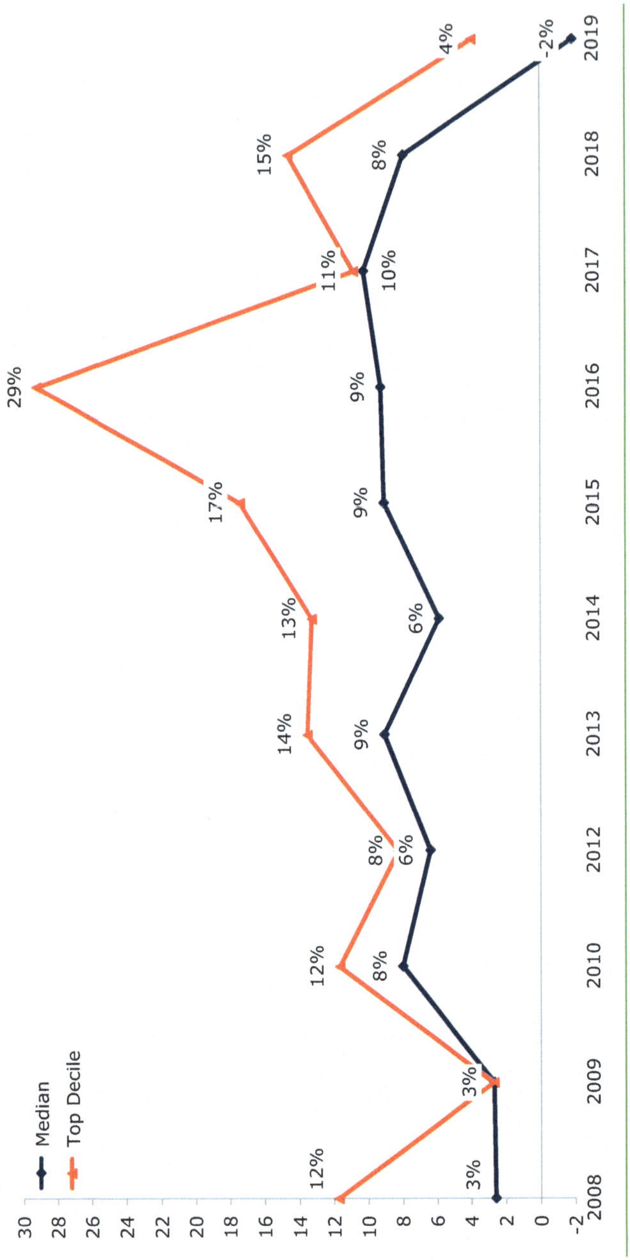

Source: Pitchbook, AlixPartners analysis

AlixPartners

Table 7.3 RE opportunistic funds are better placed in crisis time

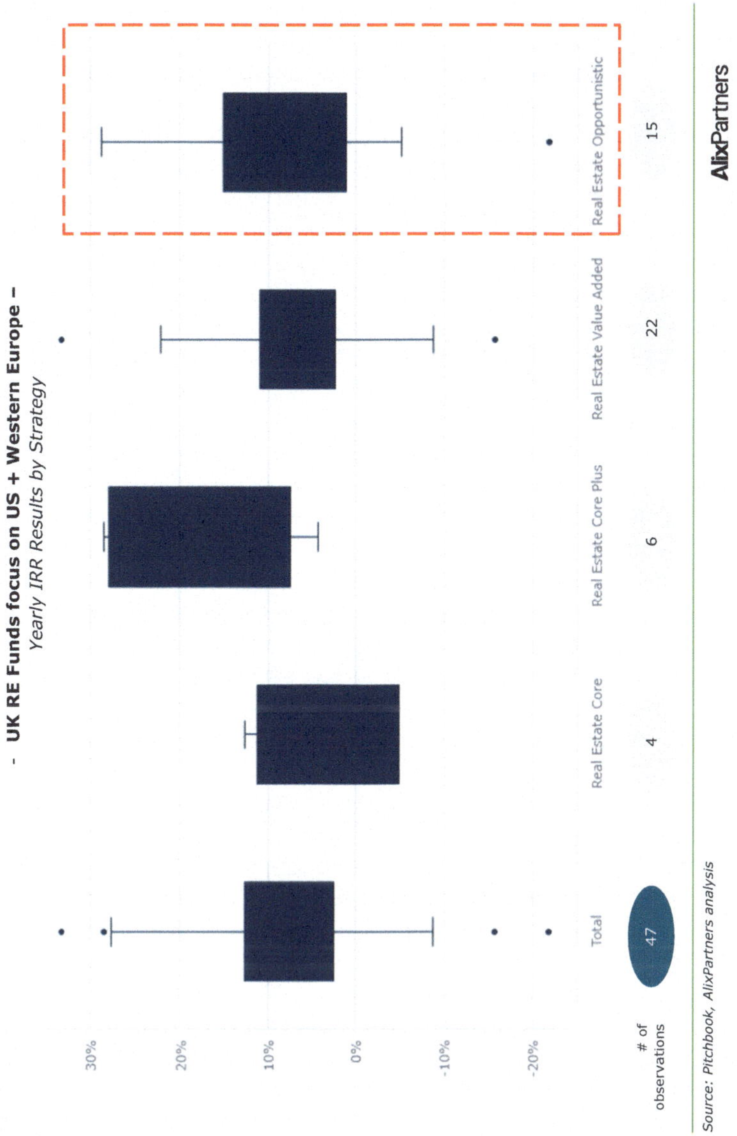

Source: Pitchbook, AlixPartners analysis

Moreover, this time, with coronavirus, it has been said, the cycle is truly different as no major asset bubble had been spotted in real estate, banks were in a much stronger capital position vis-à-vis 2008, and fiscal and monetary policies were more or less stretched and mischievous, but not abnormally so. In fact, the virus has been hitting like the proverbial comet did with dinosaurs, as an exogenous factors that was maybe badly managed but anyway almost impossible to avoid, let alone predict. If this is the case (that was the reasoning at the beginning of the pandemics), the economy and society will just be kept hibernated for a while, and then restart to life, as much as before. Alas, we have seen that changes in the world economy and society will, in fact, be much more profound and long term (Wolf 2020). Most recent estimates from the IMF are projecting a loss in GDP for 2020 for around 8% at global level, versus the near zero in 2009, just after the global financial crisis. Also, notwithstanding the anticipation of a workable vaccine for COVID-19, return to full normality could take as much as 2025 in certain countries, starting from Italy. As a result, the European banking sector is predicted to potentially face a new wave of NPL of up to 30–40% of its assets, with over-the-cycle net losses of around € 1,4 Tn (Scardovi 2020) that will require recapitalizations, mergers, and (potentially) some liquidation (corporates are also expected to have their equity base wiped out for a good third, as an average across all sectors). Eventually, even if this time the "how it all started" was truly different, will it be different anyway, when the crisis will unfold at full speed in the economy? Would have made any significant and relevant difference for the dinosaurs to be hit by an exogenous comet instead of an endogenous nuclear bomb? What matters is, eventually, the way out towards a new future.

In fact, real estate is one of the most cyclical sector in the economy. It usually thrives when things go well and GDP is growing fast, as net available income is also doing so, and lending supply as well (with the real estate stock on the supply-side being almost constant and not able to change that fast). It also dives similarly fast when things go bad, and people lose their job or see their invested savings (and perceived financial wealth) dwell fast. We could also argue that this cyclicality is exacerbated by real estate itself. As the largest asset class is so heavily impacted by changes in net available income and lending supply, its subsequent loss of value creates other chokes and constraints in both, hence supporting further crashes in the economy and then back again in the real estate sector. Specifically, we have argued, the destiny of financial services and real estate is so closely intertwined that it is almost impossible for banks to do well when the real estate sector go south (it is not impossible however to do badly when it goes north). And in turn, it is almost impossible for real estate, as an investable asset class, to do well in terms of financial performance, when banks struggle, as it is so much leveraged and relying on a demand side that is often sustained by a 70–80% of debt that is granted on top of the "highly collateralized" credit risk undertaken. In short, the starting point of this (BC) may indeed be different, but the destination point may not. If the European banking sector will suffer greatly (AC), so could do the real estate one as well, no matter what the fiscal and monetary policies that could be considered for the short term and (more importantly) could be considered sustainable for the long one and with lots of volatility in between. As it is often said, the housing market acts as the canary in a coal mine, during times of economic stress prices tend to fall as a wider economic downturn looms. Why is then, in the midst of a deep global recession, that property valuations, mostly in the residential space, have kept rising in so many countries, as shown in Table 7.4? And with wide variations across specialty sectors, but also showing some good performances as well—as indicated in Tables 7.5 and 7.6.

In fact, house prices growth has accelerated at around 4% in 2020 for OECD countries and even faster in Europe and the United States (Arnold 2020). But whether this is going to be sustainable is an entirely different question—as stated by UBS economist Matthias Holzhey, co-author of the UBS annual global real estate bubble index. In fact, stimulus packages from government and central banks have supported struggling companies and furlough employees and with a lending supply that is again playing with 100% loan to value mortgages (and with borrowing costs near record lows). If the real estate cycle is again living on steroids, it could help it survive and even surf the pandemic-related crisis, but all this could just point to a more brutal correction to come in the future and, more importantly, to a deceleration in the rate of change that is required to transform the sector and the overall asset class towards a sustainably higher level of productivity. A bad case scenario would include a sudden tightening of the mortgage offering from banks and the progressive softening or crash of the residential market, then driving down consumption and in turn investments. And an even worst case could imply that, following this, there would be huge delays and constraints in the structural transformation program that global cities were just envisioning and planning before COVID-19, to turn smart and cyber, meta, and even more "cy-phy" to become more attractive and sustainable.

7.3 The Journey and the Destination

No matter what the new steady state will be, it will be significantly different from the starting point—this the widely shared view and with new trends and paradigms emerging as shown in Table 7.7. But then, aside from the overall level of the sector in a 3–5-year timeframe, it will be important to assess who has gained from the disruption and how much has lost and how much value was migrated across sub-segments of the market, specific designs and setups, and so on. Also, it would be relevant to understand the direction of change, but also the acceleration and relative speed and any deviations in between as it could be expected that there will be many value migrations, in different and often contradictory directions over time, before reaching the "other normal." To better understand all this, big data and unstructured data specifically, captured by IoT, and applied analytics based on AI can help us in assessing not just what the final destination could be, but also the pace of change, the swings along the full journey, and so on. Much, in fact, will depend on social behavior and on how people will react to the new situation by changing their ways of working and living, at the different stages of the health, economic, and social crisis. Big data and AI analytics were already the new-new thing in real estate

Table 7.4 Good RE performance AC mainly driven by three key factors

Main factors affecting current RE trends		Case Example: Housing Market	
Monetary Policy	• Central banks have cut policy rates by 2% on average in 2020 which implies: • Lower costs for borrowing, thus further increasing yields on RE • Lower returns from bonds, thus pushing investors to look for alternative and opportunistic investments (incl RE)	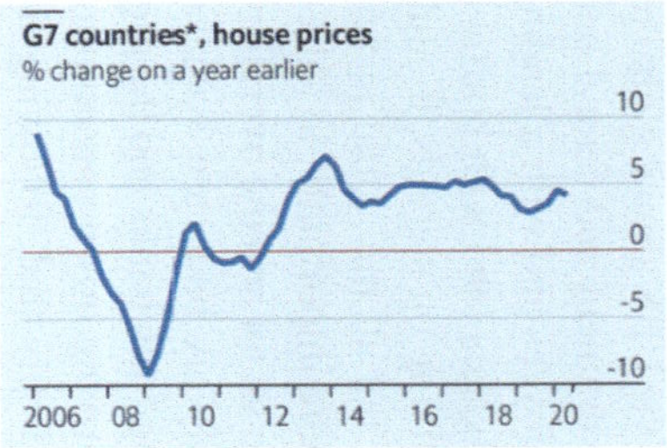	• Housing prices kept growing in 2020, with Q2 recording a +5% on average for G7 economies
Fiscal Policy	• Fiscal policy across all Western economies has been extremely lax: • Wage subsidies and direct handouts have been offered across the board, avoiding shoring of expenses • Furlough schemes have been commonly used to prevent job losses		
Buyers' changing preferences	• Change in buyers' habits with reduced travel possibilities and widespread smart working are determining : • Drop in hospitality, office and retail assets • Increase in housing prices, with buyers more willing to invest in housing, and data centers due to digitalization push		• Developers and property traders' shares lost up to a quarter in Q1, but have since recovered a large portion of the loss

Source: The Economist, AlixPartners analysis

AlixPartners

Table 7.5 More heterogeneous performances at RE sub-sector level

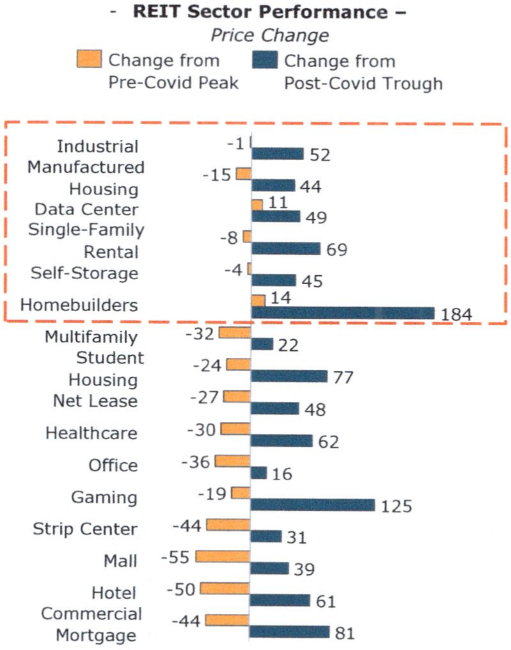

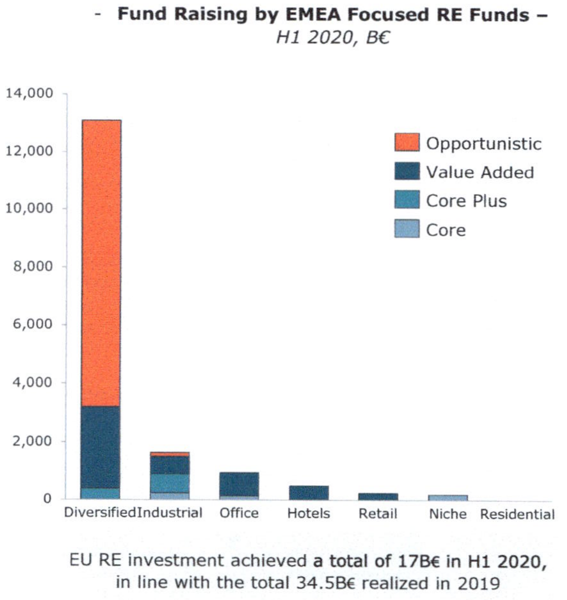

Source: Eastdil Secured, CBRE, AlixPartners analysis

AlixPartners

Table 7.6 COVID-19 pandemic has had a differentiated impact on real estate segments depending on the different asset classes

Segment	Main impacts	Outlook by asset class	
Office	• Reductions in headcount has tempered demand in the short-term (particularly as employment and wage support schemes taper) • Technology and knowledge-based sectors better positioned to drive the recovery	• Co-working • Corporate headquarter • Small office	Negative Stable Stable
Retail	• Prolonged closure of non-essential retailers has led to an acceleration of online shopping and an increased preference for local shops • Weak economy and uncertain job market has sustained consumers' cautious spending	• District market • High-street shop • Traditional shop • Shopping center • Food court	Resilient Stable Stable Negative Negative
Industrial & Logistics	• Spike in e-commerce sales has counter-balanced the slowdown of traditional economic sectors leading to a positive net effect • Increased pressure on supply chains has unveiled technological obsolescence and risk of shortages	• Warehouse • Last-mile warehouse • Data Center • Artisan workshop	Resilient Resilient Resilient Stable
Hospitality	• Stake-backed support has partially shielded hospitality RE owners from the full effects of Covid-19 • Prolonged decline in cross-border tourism and low occupancy have increased financial pressure on small/ non-diversified owners	• Residence • Bed & Breakfast • Hotel • Resort • Hostel	Stable Negative Negative Negative Negative
Residential	• Platforms deriving income from corporate workers and longer-term residents have been better protected than those oriented toward more mobile, short-term residents • Decline of international students and tourism inflows have affected student living and short rentals housing	• Multifamility • Residential house • Student living • Short rentals house	Resilient Resilient Stable Negative

Source: CBRE; Nomisma; AlixPartners analysis.

✓ Resilient ✓ Stable ✗ Negative

AlixPartners

Table 7.7 As the pandemic unfolds, new trends and paradigms are emerging and are expected to re-shape the future of real estate

Segment	Emerging trends
Office	• Occupier demand shifting increasingly towards tech-enabled "smart" space, offering development or refurbishing opportunities • Health, wellbeing and sustainability becoming the touchstones of occupier choice uncovering an opportunity early mover investors
Retail	• Enhanced focus for retail operators on preserving cash will likely translate in prices pressures, rise of flexible lease terms and attention to the full package of costs related to a location • As the structural change in the retail market accelerates, greater emphasis will be placed on the shift toward a flexible omni-channel retail model and sustainable fulfilment
Industrial & Logistics	• Occupier demand going forward to be focused on: – Enhancing supply chain resilience to avoid future shortages – Increasing closeness to customers • Accelerated penetration of digital services and new technologies driving new investments for Data Centers and IT infrastructures
Hospitality	• Ability to limit contact between customers and hygiene protocols will be the core of the "new normal" standards for accommodation, with professionally-run lodging being the better positioned • Domestic, drivable locations may grow in importance potentially leading to a "reset" of competition • Increased financial pressure on small/ non-diversified owners might force sales
Residential	• Resilience shown during the pandemic (which has confirmed the counter-cyclical nature of the segment and its ability to generate stable cashflows) coupled with socio-demographic trends will sustain long term demand • Increased focus on geographical diversification favoring non-core and capital regions with socio-demographic development potential

Source: CBRE; Nomisma; AlixPartners analysis.

AlixPartners

valuation and investing in the BC world. They were helpful in spotting emerging trends, weak forces at work, and all kind of leading indicators that, starting from the observation of the human behavior in the city, were good predictors of value migrations to come—anticipating, for example, a place becoming "cooler" or just "passé" or new ways of living a real estate that could be incorporated in its redesign or redevelopment. In an AC world, absolute and relative volatility could be higher, for a period of time, given the structural realignment required and the uncertainty regarding what would these be. It then becomes even more important to be able to couple more fundamental-based, valuation approaches driven by discounted cash flow and the analysis of macroeconomic factors driving supply and demand, to other, more innovative approaches, based on behavioral analysis performed via big data and AI.

Big data and AI would then be relevant for determining real estate successful strategies: not just because they would help in selecting them, but also because they would help in driving the competitiveness, attractiveness, and ability to dynamically readjust through time and hence appreciate in value over the long term. They would drive value through the digitization of the real estate and of the built environment. Real estate, one of the most "tangible" investable assets, was already more and more driven by the intangible part of its value BC, its link to its "digitization" and the embedding of digital, intelligent systems able to increase its convenience, productivity, usefulness given different use cases, and end users. In an AC world, this intangibles' driver of value will rapidly gain in importance as new safety standards, higher use of connectivity, and virtualization to reproduce meetings in business and social settings. From "smart" homes to buildings, neighborhoods, and cities, the "smart" (digitally connected and with a number of information-intelligence services on basic, horizontal utilities) will become almost a prerequisite for the tradability of the asset, without suffering deep discounts associated with its being disconnected from an increasingly cybernetic world. Also in this case, while clarity on the end destination (on the "smart" qualifier as well as the "sustainable" winner) would be important, also the same journey and the path followed to get there will be critical. This part will likely involve the regeneration of the urban landscape, the restructuring the buildings, and the transformation of the way real estate assets can be invested and, most importantly, used for working and living. All this will make the difference, with a number of visionary programs failing in the process and short-term pragmatic ones that will be unable to meet the expectations and needs of stakeholders in the long run and hence loosing relevance in a short while.

7.4 Financialization Ahead

As real estate value will become more and more driven by intangibles (notably, its digitization among them) and optimized by its flexibility in the way it will be able to react and adapt in an AC world and to its many challenges and then redesign itself and its multiple end uses, it will become more agile, liquid, and "mobile." With this, we are not implying that buildings or flats should become geographically mobile and ready to move across the region as some prefabricated, detached houses were used to do in the United States, something of a fashion in vogue a few decades ago. Mobile would, in fact, refer to the tradable nature of real estate. That the globally largest investable asset class is de facto very illiquid, with little transparency on market prices, long time to sell and buy, and high intermediation costs (up to 4–7% in Europe, compared with a few basis points for equities and bonds traded on regulated stock markets), does not bode well for the overall optimization of its value. It is also not conducive to the maximization of its related productivity (exemplified by the yield attainable per square meter, valued on the basis of the initial investment or cost to build). In order to optimize real estate investment performances in a world characterized by a state of flux, even more agility and flexibility in the approach is required. This, luckily, could be easily attained if real estate assets were more and better "financialized," with their ownerships pooled together on specialized (but also largely diversified) investment funds that could be traded on the stock market. We could remain owner of our flat, but for a tiny fraction only, and become so for many others, maybe in the same city or else, likely with different sizes and relative values and use cases. We could also acquire the flexibility to flip properties, with regard to our end use, if we need to move cities because of work or change flat because of our dynamic family life and structure. We could move to a larger home, if we have a couple of months ahead requiring smart working from home or settle for a smaller one, in a warmer region, if we want to spend a few months away from the buzz of the cities, to prepare our doctoral thesis or spend a second honeymoon with our wife.

Real estate funds, or even securitization vehicles, are not new and have been available for a while. Real estate derivatives (that would complement and complete well the tradable "spot" market for real estate financialized assets) are in theory quite possible, even if a few attempts have failed to build the required liquidity and transparency in the market and have been of a limited use given the lack of a real estate credible mark to market valuation. Also, quota shares developments have been tried in the past, even if with other, more limited use cases (e.g., holiday homes that allow quota holders to change holiday location every now and then or to use a single co-owned property only a given number of weeks per year). Certain markets are more developed on "financialization" as well—take Germany versus Italy, for example, with the former investing more in financialized real estate funds and the latter still mostly holding on a single asset full ownership. But the required scale of change that would be needed to optimize a € 200 Tn asset class would be paramount, opening up large opportunities in value creation for investing, trading, serving, and managing these real estate financial products. This financialization would also support the ability of the sector to react and transform in a BC world and in anticipation of a climate changed (or impaired) one. This AC financialization could then be applied to all kind of real estate and supporting infrastructure assets, by specific sub-sectors, geographical areas, strategy of investment and valorization, and cities.

We could all become shareholders of Milan, or Beirut, as the first needs to fully digitize and transform in a sustainable city and the second needs to rebuild itself, after the huge explosion and to overcome the basic utilities and infrastructure constraints. We could all become shareholders of our (or other) cities if large funds representing pools of big regeneration and development programs, core assets in need of value added asset management strategies, and even connectivity and digitization programs were available—we could take a view and contribute to the plan to make Milan smart and then sustainable, ideally investing money where our heart, civic conscience, and pursued well-being are. It would be particularly critical now, as we seek to redefine city centers, with new functions including cultural activities, bike lines, and extended outdoor spaces, redesigning the use of the space between macro geographies (cities, towns, and rural areas) and within micro geographies (cities and their hinterlands). A real estate that is financialized and mobile could also support the redevelopment of existing stocks, to reduce transportation-induced pollution, density of bodies, and social contacts and allow a more distributed use of the built environment—see, for example, the pursuit of the 15 min walk cities in Paris and Milan and Barcelona's superblocks (self-sufficient neighborhoods) or even vertical buildings (containing a small universe per se, where we can find everything, from work to romance, from leisure to health treatments, at the tap of a button, of an elevator). In fact, if cities are the most advanced laboratories for experimentation, as stated by Francesca Bria, senior advisor on digital cities to the United Nations, they can test and drive structural transformations such as the green transition, digitization, and addressing inequality and social cohesion (Sandbu 2020)—and their financialization can help the change that they need to execute.

Ultimately, huge financial resources are needed to execute such huge transformations, on top of policy making decisions and information and training programs that can drive massive cultural and behavioral shifts. Given the current state of public finances in most developed and developing economies, following the large deficits that have been run freely in response to the pandemics, and the likely bad state of lenders also because of the long-term impact of the pandemics on borrowers (including in the real estate sector), new approaches need to be experimented to get the huge financial resources that are needed to redesign and create sustainable cities. The main opportunity comes from capital markets that are able to tap into the financial savings of billion of people, private and affluent, but also mass market clients. For this to happen, real estate needs to financialize itself, getting easily tradable and mobile, as if it were a financial asset as many others. The urgency for this would be almost unprecedented, and, this time, big data and AI applied analytics could help the process, mostly by creating the intelligence to assess, monitor, and predict the changes in real estate value on an almost real-time basis. Big data can now make it possible to understand the value drivers on this most unmovable of asset classes, to get to its mark to market and related risk/return profile, supporting its transformation into a "movable" one. Along with the unprecedented crisis is then coming an unprecedented opportunity that goes well beyond COVID-19.

7.5 A Geopolitical Play

In a BC world, most geopolitical battles were already fought at the level of cities, with Paris pitched against London and Singapore against Hong Kong (respectively, e.g., as financial hub for Europa and Asia). The global attractiveness of big metropolis was then largely driving trade flows and savings, university students and wealthy residents, and so on. We have discussed how this geopolitical battles were played, in a BC world, at the intersection of finance, real estate, and technology, with a just in time approach: everything should be made possible and attractive in the city of reference, leveraging a number of globally integrated value chains that could deliver anything from everywhere in no time to every citizen, given the high mobilization reached through international commerce. In an AC world, all this should also be valid, but with the competition between geopolitical blocks left less to market forces and leading more to potential blockades (e.g., the one between the United States and the western world and China on, e.g., 5G). Even within geopolitical blocks, a reverse trend in globalization could lead to more localized competition (also favored, in the short term, by travel restrictions driven by the pandemic and the reduced international travel that is likely to remain for some time, even after the health crisis is solved). More importantly, as discussed, in an AC world cities would need to become smarter but also more sustainable, with greater resilience built-in, inside out as opposed to the outside-in previous BC approach. From "just in time" to "just in case," cities will need to build greater inventories to ensure self-sufficiency in case or emergency situations that could include pandemics, climate change, trade and military wars, act of terrorism, and everything else that could disrupt the global supply chains. Apart from better logistics and bigger inventories, this could bring back to the cities manufacturing plants and even agriculture and vertical farming, to ensure survival through times of crisis.

In an AC world, successful investing in real estate will still be driven by the interplay of finance, real estate, and technology but likely requiring a larger and larger role for big data and AI applications to guide the optimal redesign and execution of well-being-oriented strategies. Eventually, not only we could analyze safety and health, physical wellness, and psychological well-being along a continuum, with "just in case" inventories and logistic centers or re-internalized manufacturing and agricultural plants providing more security and resilience. But also, we could address them in terms of continuous digital monitoring and diagnostic and with the proposition of services which are both "just in case" and "just in time" (if you have a critical heart failure, having a local, specialized hospital is not enough, if its mobilization cannot happen quickly). In an AC world, a significant wave of redesign and reconstruction for real estate and within cities should then happen, to allow for different use cases and changes in the way we live. Real estate investing could then address the opportunities to increase resilience and build self-sufficiency at local level. Also, real estate investing in an AC world should consider the massive state-backed, Keynesian-style investments that have already been announced at geopolitical level (e.g., the New Generations Fund for Europe). They should address the market failures, the public goods, and the support of the many PPP (Public-Private Partnership) initiatives that would be required to start the engine of transformation of cities in an AC world. They should not be limiting to that but addressing as well the green conversion and other sustainability topics often related to ESG (environment, social, governance) objectives such as social inclusion and inclusiveness in the city, universal coverage of basic utilities, fairness in the access opportunities, social mobility, and so on. No matter what their final scope will be, they will play a halo effect, making parallel investments, with private capital more interesting and less risky. Eventually, building a bridge (or a tunnel) to connect Sicily with the Italian peninsula could turn out to be, from a financial point of view, a good or bad idea. But building a hotel, a few restaurants, and a shopping center in the vicinity of the access ways to them could be, regardless of that, a good idea.

7.6 The Role of the State in Real Estate BC

The State, through its government and public finance support, has always played a big role in real estate and in the build-up and development of cities, even more so after periods of severe distress and crisis, such as after World War II. Its support to (or meddling with) real estate development after World War II has been felt in a number of ways across the globe, with a few main objectives being pursued, multifaceted, but also closely intertwined. For a start, the State has historically supported the re-construction, development, and regeneration of main metropolitan cities and of large urban areas as a key driver of the country' competitiveness across borders and of its ability to grow the wealth and well-being of its citizens. This support has been aimed specifically to "special zones" of strategic relevance, usually linked to specific industries or public sectors such as health, education, tourism, commerce, and so on. The support of the State to real estate and cities has also been used to show its civic commitment to the country's society and its peaceful stability that is usually closely associated with home ownership and with affordable housing (affordable rents) for the less wealthy citizens. Social housing programs have been used specifically across the globe to facilitate the peaceful and positive co-existence of blue-collar, white-collar, and wealthy private individuals, given trends in cities' gentrification and local and global migration—a short international selection of best practice case studies is presented in Tables 7.8, 7.9, and 7.10.

An active role of the State in the real estate sector and in the market of cities has also been more proactively pursuing the overall optimization of a key component of the wealth of nations: globally, residential real estate, as an asset class, is worth around $ 200 Tn and is much more than the sum of all equity, bond, and commodity assets. Also, for most countries, real estate stands as the larger share of the private wealth owned by families, driving their happiness, perceived wealth, and hence consumption—it is then a key target considered by policy makers when managing the political elections cycle (a government approaching election day with a residential real estate market on a downward trend has less chances of being re-elected).

Table 7.8 Social housing to drive cities inclusion and sustainability (1/3)

City	Strategy	Key stats
Copenhagen, Denmark	• Pooling of public land assets (initial aim was the funding of the construction of the Metro line) • Housing cooperatives (sharing of common facilities) culture • Land-value captured through land tax, based on market price of land revalued every two years	• 20% affordable housing • 25% of new housing required to be affordable • 53% of homes privately owned
Amersfoort, Netherlands	• Local authorities incentivized to provide serviced sites for developers • Government grants subsidies to cover land acquisition, decontamination and public transport and infrastructure costs • Housing associations function according to the revolving fund principle: income gained from letting and selling homes is covers their reinvestment in new affordable housing • Housing associations receive no direct subsidies but benefit from government-backed loans	• 30% homes designed as social-rented
Montpellier, France	• Use of public-private SPVs (e.g. SERM) to maximise social and economic returns • Any new development requires 1/3 for market sale, 1/3 affordable though a subsidy and 1/3 social	• Population doubled since 1960's • +2,500 homes built per year
Zurich, Switzerland	• Central role played by cooperatives, that receive strong support from the City Council in finding land and organizing themselves • A third of housing should be affordable or "cost-price" by policy • 20% of units are subsidized to allow for those on welfare benefits	• 90% of population rents • 40k cooperative units (1/4 total units)

Sources: Link 1

AlixPartners

Table 7.9 Social housing to drive cities inclusion and sustainability (2/3)

City	Strategy	Key stats
Singapore	• State land ownership and development • Residents as «owner-occupiers» through long-term lease at subsidized prices • Funding mechanism through compulsory saving scheme	• 80% social housing • 1m housing units built since 1961
Berlin, Germany	• Supply-side subsidies for developers • Regulations to protect tenants from price increases • Partnership with public housing corporations	• Goal of 20,000 units a year, of which 5,000 rent-restricted
Vienna, Austria	• Supply-side subsidies for developers • Historical government land ownership	• 60%+ of citizens live in social housing • 440,000 social houses
Kuala Lumpur, Malaysia	• Quota of affordable houses in development projects • Incentives for the projects on private land	• 56,000 affordable units built in 2018
Montreal, Canada	• Partnership with developers (quota of affordable housing in exchange for amendments to certain regulations)	• 6,500 units developed

Sources: see case studies slides

AlixPartners

Table 7.10 Social housing to drive cities inclusion and sustainability (3/3)

City	Strategy	Key stats
US	• Federally-backed companies Fannie Mae & Freddie Mac provide liquidity, stability and affordability to the US real estate market by buying and guaranteeing mortgages issued through lenders • Lenders can then use the cash raised by selling the mortgages to engage in future lending • By packing mortgages and selling them to investors in the secondary market, they expand the pool of funds available for housing, which lowers interest rates paid by borrowers • The agencies played a large role in the real estate bubble, ended with the 2008 financial crisis, where they reported huge losses, entered in liquidity crisis and saw their stock value drop by 90% • Today are shareholder-owned companies operating under congressional charter. They continue to exercise their social function but with a more conservative approach	• Half of outstanding mortgage debt in America in Fannie Mae & Freddie Mac's portfolio
Italy	• Social Housing Bond launched in February 2020 by CDP, a financial institution controlled by the Italian Ministry of Economics and Finance, with the objective of supporting social housing initiatives in Italy • 750m€ to be used to finance construction, renovation or upgrade of social housing	• Record demand of 5B€ from 270 investors, 65% international

Sources: for US, Link, Link 2, Link 3; for Italy, Link 1

AlixPartners

Internationally, a number of schemes have been developed by States to get the government and its public finance arms support real estate and cities, with specific reference to the housing component (residential assets meant for ownership or rent by families). These have included financial guarantees offered by the State to retail (e.g., families) or corporate (e.g., real estate developers), but also real estate mortgages at subsidized prices and often underwritten at a loss (e.g., Fannie Mae, Freddie Mac in the United States). They have included tax credits extended on the purchase or rent of real estate by families or for its refurbishment and upgrade to higher standards (e.g., Grade A, "green" certification, safety standards, improvement, and enlargement (e.g., recent post COVID-19 program in Italy). The State intervention has sometimes extended to regulated pricing on new homes and rents as required quota in new developments or for entire neighborhoods at city or country level, to ensure the targeted supply of affordable homes, given demand and trends in gentrification and migration across regions of less wealthy residents (from less developed countries or from rural areas).

7.7 The Role of the State in Real Estate AC

If the role of the State was important in BC era, it is going to become even more so in an AC world. In fact, not only real estate and cities are seen as a key driver for the reconstruction and recovery of impacted economies, but they are also seen as requiring even more of a radical redesign, redevelopment, or regeneration, to make them more "sustainable," at the light of the current and future crisis. Eventually, the crisis could turn to be a catalyst and accelerator for the transformation that was already required BC. Some governments and local municipalities are starting to think not just of doing more, but also to do this more effectively and efficiently, given the increased constraint on public finance posed by the pandemics, e.g., with a greater and greater recourse to private finance. In some instances, they are trying to involve institutional and also retail investors, usually through capital markets and with some kind of a financial institution setup (e.g., as a bank or as an asset manager or as a broker dealer or insurance company). They are also considering doing this through regulated financial products (e.g., with the help of real estate or credit to real estate funds, securitization vehicles, financial holding structures with eternal capital, and so on). Overall, they are trying to pursue not only their main objective of supporting affordable, competitive, and compelling housing to their citizens but also to potentially address the broader range of objectives associated with the reconstruction and transformation of their country. For example, a financial institution license or the setup of specifically regulated special purpose vehicles could allow to governments to raise third party money through bonds issuance, securitization tranches, mutual fund "funnels," equity crowd funding, or even deposit gathering. Specifically, a State backed fund, even if managed by professional third parties, could consider a wide range of options, including becoming a bank, able to then gather deposits online or via physical branches (e.g., leveraging the Post Office network or as a wholesale bank) through certificates, term deposits, and bonds and any other money market or savings products that could be sold by retail and commercial banks in the country.

It could also become an insurance company or a pension fund that can raise and manage a "float" of money (the premia paid by policy holders that have an inverted cash flow cycle vis-à-vis the one of real estate long-term mortgages), as investing in long-term real estate mortgages is also a natural hedge to their typical liabilities. Alternatively, it could become an asset and investment manager, able to structure a number of targeted investment vehicles and solutions (e.g., real estate contribution-based, multi-seller funds; equity or debt real estate funds, REIT (Real Estate Investment Trusts), etc.) that could raise money from private institutions and retail investors alike and manage their savings optimally. In an even more ambitious and holistic vision, a business model of a real estate wealth-being (wealth and well-being) manager could be pursued and translated into a more detailed and granular business and operating model and with a mix of setups and financial product offerings that could include most of the abovementioned propositions. A number of structured solutions could then also be at play and even for application to government-owned real estate assets. For example, Dag Detter, who has led the restructuring of the Swedish portfolios of public assets, has recently argued and proposed the formation of Public Wealth Funds (PWF) that are wholly publicly owned but politically independent holding companies operating with strict commercial corporate governance standards and for the purpose of maximizing value for the public sector and thereby adding revenues to the government beyond taxes. In fact, governments and local authorities own a vast amount of commercial assets. These real estate holdings alone could have a value equivalent to the gross product of the local area, if accounted at market value. And if professionally managed with a view of optimizing their productivity, yield, and overall attractiveness, they could generate additional revenues that would be enough to cover some of the gap inflicted by coronavirus (Detter estimates an incremental contribution to GDP, as additional revenues to the public sector, of as much as 3% for the United Kingdom) (Detter 2020). They would also maximize the value of the assets.

Their idea of an Urban Wealth Fund (UWF) could actually be further developed as public funds could then leverage private, retail funds and be managed by independent, private, and highly reputed professional parties. With such a setup, independent investment management companies could then gather further capital from pension funds and insurance companies, requiring long-term and reliable returns—pursuing affordable housing but also "smart" buildings and, more importantly, sustainable cities. The urban regeneration and re-development could actually start at the level of the real estate assets already owned by the State—optimizing their use: maximizing their yield if trophy assets (as rented to prime clients at prime, market rates, as a good substitute of taxes) or their social impact (for less trophy ones, efficiently and safely transformed into digitally connected and inclusive affordable housing). An UWF, as independent and professionally managed vehicle, would also be the best place to consolidate all the historical, artistic, and speculative real estate assets that local governments have inherited recently. If we just consider

Rome or Milan, there are uptown, artistic, and beautiful places that are used for barracks or prisons. And other very expensive (but with little artistic value) more speculative assets—for example, as repossessed from criminal organizations that were using them to launder money or even host prosecuted criminals at large. The collapse in commercial real estate yields in an AC world could also create acute financial problems if these assets were owned by local authorities, if their poor utilization and lower yield are not properly addressed for transformation by independently and professionally managed vehicles. Ultimately, the health-related crisis has shown the fragility and weak fundamentals of many parts of our economic fabric. It could as well help in uncovering the hidden strength in the real estate and infrastructure assets sitting, mostly idly, in the balance sheet of central and local governments and families as well.

7.8 A Private Equity Opportunity for Everybody

Real estate, even if (or maybe because is so) little financialized, has been targeted by private equity investments for decades, with large players managing hundreds of billions of money and with huge "powder capital" ready to be deployed. Performances have varied a lot, but overall returns have been good, also helped by the relative illiquidity and limited transparency on the market dynamics. As such, it has been an opportunity almost exclusively tapped by institutional investors and a few professional clients (i.e., very wealthy individuals that can invest on single ticket north of € 500K). In a BC world, several regulations have limited, across most jurisdiction, the direct investments of retail clients (including affluent and mass market citizens) in this asset class and with a private equity, more active and speculative approach. In an AC world, the once-in-a-lifetime opportunity is to do just that. On one side, the sheer amount of equity wipeouts on the economy caused by the pandemic has been leading to a huge re-leveraging of the economy, either through public debt or lender-based private one. This will then require a rebalancing to sustainability that can only happen if real equity is reinjected in any kind of real asset in the economy, starting from real estate and, more broadly infrastructure ones, mostly in the context of urban areas and metropolitan cities. On the other side, the higher cost of capital required by institutional, more speculative investors would make most of these investments unable to repay their loans, unless a large portion of these were written off. They would also fit less with a longer-term investment thesis that aims to create value through a profound transformation of real estate assets and entire areas, by making them more digital, green, and sustainable. Certainly, opportunities for tactical investments ("special situations" performed by "opportunistic funds" in the professional jargon) are not lacking in any way, for the benefit of the private equity industry itself, also given the expected higher level of volatility AC.

But in order to create an opportunity for real estate, as a broader sector, and for cities to redesign and transform in an AC world, another kind of private equity needs to be considered and put at work, with more moderate targeted risk/return trade-offs and a longer holding period for investment. Cognizant of the equity gap created by the pandemic, policy makers around the globe are now considering fiscal incentives and a changed regulatory setup, to support the investment by retail individuals in the real assets of their country and as private equity—professionally managed by investment management companies. Past similar experiences, through REIT (Real Estate Investment Trusts and other funds) have been mixed, but private equity-like opportunities for investments have actually never been available for people. Such an opportunity, made consistent with the current MIFID 2 and the up-coming MIFID 3 regulatory setup, and sold and managed with appropriate standards of transparency, advisory, and liquidity, would form the basis of a very much augmented "participation" of citizens in the reimagining, redesign, reconstruction, and transformation of cities, starting from their real estate and infrastructure assets. Not only that. Such a "capitalism for everybody," as suggested by Franklin D. Roosevelt in 1936, would just stand as the right support needed for the economy to recover and rebuild stronger than before and as a good chance for affluent and mass market citizens to engage in a value adding diversification of their investment portfolios, with less bonds and equities, less money market products—all subject to inflationary threats—and greater, direct exposure to real assets that they can see, touch, and use as normal citizens. Private equity in real estate in an AC world would then become an opportunity for everybody and for cities that would have a direct stake in them and a better civic conscience.

7.9 From the 15 Minute City to the Integrated Building

In an AC and climate change affected world, the shape and form of the city will also need to change, with polycentric trends and approaches to redesign the city, most of them already in place BC but now accelerating and extending big times. A metropolis such as Milan was already planning and developing new urban areas as centers of attractiveness—from CityLife to Milano Santa Giulia, from UpTown and Area Expo to Milano Sesto. But what about the "15 min walk" city that is becoming a new trend in a post pandemic world, where everything is placed around us, to limit virus transmission and traffic congestion? More than that, we could even think of an "integrated building" proposition, where a single, fully digital, green-compliant building is set up to be a "city-in-a-box.. Apart from horizontal utilities, such as security and first aid, or energy saving programs and tailored, just in time procurement, that could be embedded in the building competitive proposition, leveraging its IoT and big data platform, its vertical functions could also evolve, to replicate everything we need from the city, in a box. Gyms, swimming pools, and spa centers could be hosted in its basement, as already common in many international cities. But kindergartens and primary schools could also be added and a shopping center (for what is not already supplied on demand and via an e-commerce proposition that is activated by our own voice and via the pod we have installed in our flat). Temporary offices, for professionals, with augmented reality videoconference facilities and 3D printers could also be found on a dedicated floor, with several restaurants and a garden, maybe, at the top floor. Residents would not just enjoy an eHealth monitoring system connected to first aid services, but likely find in the same building a diagnostic center and a hospital by the hours (as we could be moved and monitored from our own bed, just a few meters away), with resident doctors and travelling specialists that could virtually replicate all the services offered by a big, structured, and traditional hospital. Other hyper-vertical offers could also be possible and made economic, if the level of specialization and scale required could be driven not just by the scale of the building, but also by its focused value proposition (if a 20 floors building is targeting young generations, in their 20s and 30s, it will more rapidly reach the scale required to offer some university programs than building with 40 or a 100 floors but unfocussed on the target market; the same would apply, e.g., for senior living buildings targeting elderly people or else).

In summary, in a AC world, and leveraging the power of digital technology, but also new concepts and designs related to specific and attractive value propositions, we could move from a main downtown city, to a poly-centric city, to a 15 min walk (hyper-poly) centric city to an integrated buildings city, where all key buildings stand as an almost self-sufficient new center of gravity for business and social lives. We could move around at the tap of a button (the one of fast elevators) as everything would just be a max of 15 floors away, hence reachable in a matter of a few seconds. Not just all common and more public services would be available and professionally supplied just some floors away (from the gym to the dentist), but also all the common and more private services could be centralized and provided by "The Building" (from cleaning to maintenance). All apartments would be fully serviced, as would all remaining public spaces, with higher standards of quality, a greater timeliness, and economies of scale and cost (not to mention the hassle-free proposition offered to residents). In truth, and following the financialization of the real estate as investable and more liquid asset class, we could just go renting around, moving from flat to flat according to our own demographic dynamics (as single, with a partner, with kids and a dog, then as divorced, and so on) and lifestyle (with full access to gym and swimming pool or greater help required on health monitoring and diagnostic). Instead of buying into a single, rigid, indivisible, largely unmodifiable and unmovable flat, we could become invested in a "virtual flat that follows the individual" concept, likely characterized by different service and quality level (eventually, you can live in a student's house as "bronze," then move to "silver" as just married and "gold" or "diamond" as you retire, and you could follow the level of quality on offer as you change city or neighborhoods). We could then become shareholders of a single, large building (moving floors and swapping services as needed) or quotaholders of a real estate fund comprising a set of buildings in Milan (one for students and one for young couples with children or for divorced people and then elderly ones). Not to mention that, in times of pandemic, the emergency area to address would not be a country, or a region or single city, but single buildings that would be much easier to monitor, trace, and track and quarantine (and with maybe aerial, suspended, fully monitored tubes to connect main buildings instead of streets?).

7.10 New Paradigms for Real Estate Investing

A number of trends were already in place in real estate in a BC world and requiring an adapted approach to real estate investment and management. Some of these were already implying paradigmatic changes to real estate (and cities) as an investable asset class. In short order, four emerging paradigms could have already been detected. First, the tangible "par excellence" real estate asset is becoming more immaterial, with a value that is more and more critically driven by intangibles (its digitization among them). Second, the "safe as a house" asset that had been used since time immemorial by banks to collateralize their lending risk has, in fact, emerged as a driver of very relevant risks in itself, as its value—that was assumed to be relatively stable through time and always growing over the long term—has shown to be quite volatile on a mark to market basis, if that was just widely available and with no assurance of capital gains over the longer term, if inflation and purchasing power parity adjustments are taken into account. Third, as discussed, the very unmovable real estate asset is increasingly becoming mobile, through its growing financialization and structured solutions that can allow the fast and cheap transfer of some components of its risk/return profile. Fourth, as digitization has been used as an accelerator to all of the previous three drivers of change, real estate has been encompassed by the "smart" revolution. Cities need to become smarter and smarter, and companies and people as well and their real estate included. All these trends are likely to remain in place, if not on an accelerated path, in an AC world, in the context of macro- and microeconomic scenarios that look rather grim for the short-mid-term.

Based on performance up-to-day, the real estate sector has not done badly, thanks to lax monetary and fiscal policies and the increased demand for residential assets from people that have been forced to work and live in their homes for long periods of time. But the future investment performance of real estate as a sector overall looks very uncertain, and—literally—everything could be possible. Most certainly, whatever the high uncertainty and volatility of the sector, more heterogeneous performances are already observable at the sub-segments level, with public spaces (e.g., offices, shopping, and entertainment venues) doing worse than private ones (e.g., flats) and with suburban areas doing relatively better than single-center downtowns. Also, logistics is doing great and manufacturing could be OK, but with a major downside risks for food and beverage shops, cinemas, and gyms. More broadly, large redesign and reconstruction programs should be considered to refit existing stock to new, updated purposes and develop entirely new concepts across vast areas of built environment or single buildings and progressively extending the focus of the transformation from "smart" to more holistic approaches, focused on sustainability and on the optimization of the cyber and physical dimensions. Overall, most of the changing paradigms hypothesized for a BC world would still stand, with maybe a few additions or evolutions. Specifically, a greater focus on ESG targets should make the "inclusiveness" of real estate more relevant vis-à-vis their "exclusiveness." Neighborhoods should also be judged as more sustainable if they promoted social integration and cohesion as opposed to gentrification and class stratification. And a greater fungibility of real estate assets, to support the shared and circular economy and promote the green conversion of cities, should also become more strategically relevant, along with the greater resilience offered by "just in case" capabilities. Finally, new paradigm should also emerge regarding the design of the built environment, with new concepts like the "15 min walk city" and the "integrated building."

References

Martin Arnold, Surge in European house prices stokes concerns over market resilience, Financial Times, November 6th, 2020

Dag Detter, Making real estate assets work for local governments, Financial Times, November 6th 2020

Martin Sandbu, How to make cities more liveable after Covid 19, Financial Times, 22nd November 2020

Claudio Scardovi, Analisi comparativa dei sistemi bancari Italiano, Francese e Tedesco, AlixPartners working paper, Aprile 2020

Martin Wolf, See for example "Ten ways coronavirus crisis will shape world in the long term", Financial Times, November 3rd, 2020

Chapter 8
Open Data in Open Cities

8.1 Electrifying Future

COVID-19, climate change, and the many geopolitical tensions and decoupling that could turn into trade barriers and protectionism, or even regional and global wars, look like just as the preamble of even more disruptions to come. They all share a common theme, reminding us of how weak we are and of humankind's fallibility (and even existential risk) through history. We discussed at length COVID-19, which came up as a totally unexpected event during the writing of this book. In fact, climate change looks like bringing even greater disruption in the world of cities to come—with the pandemic giving us just a preliminary idea of how hard it will be to deal with this. As stated by *The Economist*, as economy activity has stalled because of lockdown, 2020 will likely see drops in CO_2 emissions of between 4 and 7%. But in order to keep the earth warming target of less than 2 °C, CO_2 would need to fall to near zero by 2050—without halting the economies and locking down entire populations but rewiring the first and reshaping the second (The great disrupter 2020). The rewiring of the economy will require a radical redesign of the energy sector and of its global geopolitics, with powerful petro-States of the past giving room to even more powerful electro-States of the future. With clean energy gaining momentum with both investors and end users (and voters) and with interest rates near zero, more and more international politicians are now backing accelerated green-infrastructure conversion plans. The shift from fossil fuel (today accounting for 85% of the energy produced globally) will be evident everywhere and ever so much in the world's mega cities, which will have to rely on renewable and green energy to provide mobility, heat and air conditioning, and the basic support to commercial and manufacturing sectors working in their extended urban and suburban areas. Energy will also be needed to support services to the person, such as healthcare.

Indeed, a great part of the competitive endgame of global cities could depend on how fast and effectively they could move to transform into electro-cities (Power in

© The Author(s), under exclusive license to Springer Nature Switzerland AG 2021
C. Scardovi, *Sustainable Cities*, https://doi.org/10.1007/978-3-030-68438-9_8

the 21st century 2020). Their electrification could come from multiple sources, with solar panel, hydro-energy, and wind turbines among them and used to provide energy as electricity straight-off—as current and via batteries. This will also require a significant redesign of the city, from its building and infrastructure to its mobility and communication systems, to allow electricity usage to rise from around 20% in 2018 to over 50% in 2050 (Petrostate versus electro-state 2020). Cities will then need to invest strategically all along the electric value chain, leveraging better technologies for production and distribution and also designing more effective and efficient use cases for its optimal use—anything from public utilities to tailored services to businesses and to the individuals. While at geopolitical level, the game to became the most powerful electro-State will be played by the usual giants, with China in a great position to play such a role (because of its dominance on solar and wind energy, control of most "rare earth" global supply that are critical to produce batteries, and so on), main cities across the globe will also play a very relevant hand, because of the way they can influence and shape the transition and then end usage of green electricity by people. The transition towards electrification will then involve massive investments—the International Energy Agency (IEA) estimates around USD 1.2 Tn extra annual investments to rewire the power system (Petrostate versus electro-state 2020), with most of these funds spent on urban areas, to make energy use smarter and greener. As cities are facing this electrifying future, a couple of obvious considerations stand out. The first, as everything will become more and more driven by electric energy and with digitally connected systems aiming to transfer energy (e.g., to allow real time, wireless recharge of small IoT devices) and gather and share data that may be helpful in ensuring a more efficient use of it, digital transformation will follow and at the same time support electrification. Second, as digitization will allow the creation and use of cyber twins (at city, building, and home-office level) and promote a number of new, business models and way of working and living, electrification will also be supported by this transformation and contributing in turn to make the new world of cities not just smart and cyber or meta and "cy-phy," but also sustainable. New electro-cities for building a more sustainable and competitive future.

8.2 Open Data Revolution

In the new, electrified and digital, cyber and physical world of cities, data will become even more relevant, and, because of this, their open source access will likely feature as one of the most relevant demand (and challenge) of democratic societies—a critical basis to ensure inclusiveness and the equal and fair access to opportunities for development and wealth for everybody and for the technological and scientific progress to come. This would be consistent with the very origins of the world wide web that was created in 1989 by Sir Tim Berners-Lee on the basis of his very romantic vision that was aspiring to bring people together by connecting anything to anything. From there, things have changed a bit and people talks now of "digital feudalism," to describe the control big technology platforms have over data. As a result, he has co-founded a start-up, Inrupt, that aims to shift the balance of power, by putting data back in the hands of the people (Free the data serfs 2020). This data is now the world's biggest business. According to *The Economist*, $ 1.4 Trn of the combined $ 1.9 Trn market value of Alphabet and Facebook comes from users' data and the firms' mining of it. More for them could come if they could get their hands on offline data, their weakest link. Hence their interest on IoT and the many sensors that could be embedded in the smart city, from cars to kitchens that are expected to churn even more personal, unbiased information on us. On the many ways to limit the power of these potential data feudalists, the empowerment of individuals, collective actions, and the role of governments and anti-trust bodies have been mentioned most frequently (even more so after the first official enquire on Google monopolistic position was formally launched in the fall of 2020 by the American Department of Justice). In fact, a fourth way could also exist if data, anonymized when of personal nature and limited or excluded if confidential, are instead offered to everybody and for free. If data are the scarce and most productive resource in the economy, it could be a good idea to make sure they don't generate unassailable monopolies or oligopolies. But ideally without taking them off-market and instead making them (because of their inherent nature—data don't get consumed by use and, pooled together, are even more valuable) available to everybody, leaving to the best individuals and firms to get the best intelligence and use cases out of them.

As an "open data" movement (or even philosophical stance) could become one of the most prominent forces for change for markets and societies alike, an obvious question could then be from where all this could start from. In fact, an "open data" revolution may have actually been starting in one of the most unlikely places—that is, banking. Following the European Countries' directives aimed at implementing the so-called PSD2, a number of API (standard, technical protocols that are aiming at making this "open data" approach feasible among corporate counterparts, at the whim of the consumer that is the ultimate data owner) have now been implemented. On this basis, a number of specialized data managers and other FinTech have been trying to develop a competitive business model, offering easy data extraction and aggregation, or data mining and presentation and visualization of elaborated information, when not straight intelligence. Dynamic comparison tools—comparing different current account commercial conditions, for example—are now available. Other platform-based utilities are instead offering advice to perspective clients on how to save fees with banks or optimize their family's weekly budget or multi-generational planning, on the basis of the analysis of their consumption patterns, stage in life cycle, and available income.

More sophisticated offerings are also promoting "virtual CFO" solutions for small and medium enterprises, or lifestyle advice for targeted retail clients, all built around the tailored proposition of products and services that can help them in addressing the everyday issues they face or the more structural hard problems and ontological aspirations they may be trying to solve or pursuit. For all this, it is good to start from the financial data that also carry with them other relevant information (such as the time and location of a customer when doing a certain transaction, or the product bought and the vendor used in the process—available, for example, from the end user credit card information). But having further, non-financial information regarding, for example, the online behavior of individuals and corporates is also very relevant—hence the foray, at the European Union level, to eventually consider the extension of a similar, binding approach towards an "open data" for "big digital" players. Not only this would allow the more consistent pursuit of a "level playing field" of competition within and across regulated and unregulated sectors, but also, it would support the extended suffusion of financial and behavioral information. Indeed, even considering such an extension, the current "open data" revolution is very far from its full potential—which can only happen if two parallel goals are pursued. On one side, as we discussed at length, the richness and explanatory power of data would be highly augmented if online financial and behavioral, bite-based information could be complemented, mixed, and integrated with the information regarding the behavioral dimension of people and things when acting in the offline, atoms-based world.

Such data should, off course, be subjected to the same rules governing the access to the "open data" universe of the online world and respecting the privacy and ultimate ownership of the individual with regard to private information. For this first goal to be achieved, cities, as cyber-physical environments, with extended, IoT-based connectivity and AI-driven applications running on centralized and decentralized big data platforms (BDP), would need to play a critical role. On the other side, the implied potential of the info-telligence economy, and of its impacts on the overall society, would also get magnified if data, of a non-personal and confidential nature, would be made truly open and easily available. Such data openness (and hence inclusiveness), and their related use, would also need to be regulated and monitored, to ensure that a right set up of incentives on, for example, R&D and for the development of intellectual capital is maintained. For this second goal as well, the role of cities would be paramount, with their digital infrastructure and physical setup, to ensure the best availability and seamless accessibility to the open data being developed and that ideally should be put at use for the pursuit of the public good. In an ideal cyber-physical city, most talented individuals and most productive companies would be supported in their economic and social pursuits with an easy to use access to the most relevant information—ensuring therefore that maximum wealth and well-being could be created, in a sustainable and most inclusive way. This is why not only an "open data" revolution could hardly happen without the support of cities. But also, cities could hardly develop into "cy-phy" sustainable ones, without the support of "open data."

8.3 Going Soft

If banking can be considered a most unlikely place for the "open data" revolution to start, Microsoft could similarly be considered a most unlikely champion of it. The software giant from Cupertino, once touted as a by-word for a technological walled garden and then demised for an early decline, after the return of Apple and the birth of the i-Phone, has now got back on top of the most valuable companies in the world (as of November 2020). Even more unexpectedly, it has recently joined a fledgling movement to liberate world's data, with plans to launch multiple "open data" groups by 2021 and give away some of its digital information, including the one that it has gathered on COVID-19. In fact, the global pandemic may have critically boosted the efforts to promote data sharing across countries and geopolitical blocks and even between private companies (e.g., for the search of a vaccine as a global solution to the crisis). It may as well have led most people to be more relaxed about the use of their personal data (including location and health), if done in the pursuit of public health reasons and for their own safety. Climate change, widely assumed as the next "black swan" to come after the coronavirus one, may also accelerate this open data run and contribute to the creation of standard protocols for the fast, easy, and cheap access to available information on carbon emissions and extreme weather conditions. In short, the run towards more "open data" (and a more intrusive role of the managers—public or private—of these big, open data platforms) appears as a trend here to stay. As such and leaving aside the pandemic crisis and no matter what the political convenience or commercial opportunity, it needs to show and demonstrate that "open data" are good for more competitive, sustainable cities. And that these sustainable cities are better suited for the pursuit of a highly inclusive, well-balanced public good—and more apt to support the development of better economic and social systems that can make countries more successful.

In fact, according to the OECD, a club of most rich countries, it can be estimated that if data were more widely exchanged, many States would enjoy gains of 1–2.5% of GDP (Tear down this wall 2020). This is mainly because data are "non-rivalrous" and—unlike oil—can be used and re-used without being depleted, to power various AI algorithms. But also, because if more data are more widely exchanged, in transparent, liquid, efficient way, they tend to better reward their original owners. In an "open data" framework, data do not need to be exchanged for free, and, actually, if an efficient market for exchanging information is in place, consumers are more likely to get a full valuation for their data, with a price that is achieved by comparing competitive demand and supply forces. On the opposite, this is not the case, as we discussed, when consumers "cede" their information when joining a social network or search engine or emailing system, getting some "free" service in exchange for that and with tens of pages of terms and conditions that are unilaterally set as non-negotiable. Also, if data are unconstrained, they can be used multiple times and ideally by their best available users, hence optimizing the chances of innovation, growth, and value creation in the overall economy—the data "artificial "scarcity is addressed and resolved, and information are put to their best use and manipulated by the best talents to get the most valuable and serviceable intelligence out of them. In short, creative destruction is pursued and ensured, at the level of info-telligence economy, where destruction is not of the information itself, but of the special interests and risk free arbitrages that were attained by very specific constituencies, private or public, on the basis of their privileged but otherwise not justified access to such data. Information oligopolies could, in short, be swept away. Specific software for sharing and protecting data is already available, and licenses and rule-based frameworks are also being developed, to provide access to data without losing their control.

8.4 Open Cities, Open Companies

A number of global, well-renowned cities are already launching "open data" initiatives, to spur their own competitiveness and increase their overall attractiveness—from Berlin to San Francisco, from Milan to Toronto (a contextual democratic set up in the country of residence is of some help for this). Also, a number of global, well-reputed companies are launching "open data" initiatives, to increase their innovation capabilities and strengthen their extended ecosystem—with business partners and commercial clients all collaborating in an osmotic environment, from BBVA, a Spanish bank, to GlaxoSmithKline, a British drug-maker. On the same token, public and private universities are promoting the "open data" meta theme, focusing on specific venues of research and teaching. There is a critical nexus in here, between the city level initiative—typically public institutions' sponsored—to promote and foster an overall "open data" environment and regulatory framework, and the company level one, as private pursuit of "open data" on behalf of their shareholders, to spur innovation and gain some competitive advantage. The latter can operate more effectively in a supportive cyber-physical ecosystem (sometimes called "sandbox," with light regulation and open, accessible standards to allow innovation and quick testing and proofs of concept), and the former can become in turn more attractive, for well-known, mature companies and for start-ups as well—with their armies of professionals and would-be entrepreneurs following suit (along with their families). Of course, a functional mobility infrastructure, a well-designed city-dwelling and beautiful architecture, with open spaces aimed at improving the well-being proposition can help. As it would help a lower taxation or, for the innovation and development stage, some "free zone" (or "economic special zone") status that could be granted by the central administration of the host country and spread across a number of key, strategic metropolis that are aiming at global competitiveness.

It would also help, in order to attract companies, and university campuses, that are made of all-too-human people, if the host city or neighborhood has a great climate, a high-standard environmental approach and low pollution (increasingly, a key driver of choice when defining place-location strategies) and an historical heritage and arts maybe not to mention some beautiful seaside setting nearby. In the incipit of our previous book (Finance and real estate wealth-being 2020), we have been wondering why Singapore, and not Palermo, has been surging as global metropolis and attractive hub of choice for most global, fast-growing companies. We have then been arguing how Palermo, with its € 500 market value per square meter for real estate assets based in downtown, its historical heritage, artistic beauties, nice weather, and central location as seaside access for the Middle East and North Africa to the western European world, could have a great potential to became just that and for EMEA companies and start-ups. In that book, we have been arguing as a mix of finance, real estate, and technology could make the trick (given, of course a proper policy making—regulatory, fiscal, rule of law, etc.—setup). Within this framework, and in a post COVID-19 world, technology has become even more important, along with the potential access to a pool of highly relevant off- and online data—made available through easy and quick connectivity, transparently and in respect of current GDPR regulations. These data would also be supplied with tools and approaches to provide for their trading on a fair basis and with market mechanisms that would include the cheap or almost for-free access to certain sets of non-contentious, non-personal data. In fact, this "open data" would be the new oil that could transform the value that is embedded in the many beauties of the Sicilian city and make it the next Singapore-like, success story of innovative, cyber-physical, sustainable city that is able to attract capital and talents, create wealth and well-being, and position with a uniquely attractive proposition.

The challenge is certainly huge, if we take the relative starting point of Palermo and Sicily vis-à-vis Singapore on many critical dimensions and also from the digital standpoint—with a number of cyber gaps to close and overcome. On the same token, Palermo and Sicily would have a huge potential from the physical standpoint—with plenty of lands and seaside, historical and artistic beauties, and nice weather and with pools of people with innate talents and significant cost structure advantage. The development of an "open data" approach and of a consistent "cyber-physical" optimal design could just play that trick. Regardless of the fact that this is the next winning challenge for Palermo or Singapore, the development of an "open data" approach, in a well-defined "cyber-city" design and competitive proposition for the urban dwelling would also drive the development of their social and economic ecosystems. In fact, it would stand as a critical contribution to the resilience of the city in question—with relevant implications across all ESG (Environment, Social and Governance) dimensions—hence to its becoming a "sustainable city" of the future. This would be so not just because of the increased efficiency and efficacy potentially linked to all kind of open data and AI-based applications, including the ones regarding the better management of climate change risk and the greener management of the environment. But also, because the easier and greater sharing of data would counteract the concentration of the great financial wealth-being created, allowing a fairer sharing of well-being and of economic and political power. An open data setting, supported by cities and promoted by companies, would provide a better governance framework for societies and markets to operate and flatten the polarization of citizens—bridging the "data divide" (in the words of Brad Smith, President of Microsoft) and promoting an extended inclusiveness. That would be and end per se, not short of a great vision and well beyond any "smart city" ambition.

8.5 The Real Risk in Front of Us

For such a scenario of extended and deep digitalization of cities, one of the main risks for us to face would be cyber risk and mainly related to the development of IoT networks that become more and more embedded in the built environment, from its private spaces to the public ones. In a "smart home/office" setting, a number of "smart" products have become commonplace and increasingly adopted—from thermostats that adjust to home occupancy levels to wearable devices that detect health risks. Most of our products have become interconnected, even including connected toys that can be turned into convenient tools for hackers, with increasingly dangerous hazards, starting from spying and accessing and harnessing personal data or manipulating our children' minds (Venkataramakrishnan 2020). On top of being powerful networks for funneling data of a potentially critical nature, these IoT devices and systems have also the ability to adapt, through AI algorithms—then posing further risk for their improper use and manipulation through time, as manipulating them would imply manipulating our life (Green 2020) (think of an IoT that is replenishing automatically our fridge based on our consumer's preferences, with AI monitoring our diet, across food categories and even brand, and think of the consequences if those algorithms were controllable through time in an imperceptible way). The power of these IoT devices will grow as they will become increasingly embedded in the real estate we live in, actually contributing to its valorization and attractiveness: who would not want a flat with a set of robots that are cleaning up and feeding the fridge and operating the laundry based on our needs, with little or zero assistance required from us or from any other third party people? The risk of a cyberattack or fraud would then extend to such common utilities like energy, from the thermostat we have at home to the electricity grids that are used to run an entire city or even country. As power infrastructure (from home/office to city/country level) is upgraded and becomes increasingly reliant on IoT—an evolution dubbed as the Internet of Energy (IoE)—cyber criminals have more opportunities to penetrate and corrupt the system at hub or spoke level, with ransomware scenarios both at corporate and retail level and disruptions on the supply-side (Internet of Energy 2020).

Cyber risks are also on the rise in an increasingly connected city, where technology underpins the main utilities we have access to and also monitoring (and potentially manipulating or interfering with) the way we behave in public spaces. The increasing use of digital payments, for example, was already a clear trend before the pandemic and the development of smart, IoT interconnected cities. However, on one side COVID-19 has accelerated the rise of digital payments and the demise of cash. On the other side, the development of IoT networks that can monitor our buyer behavior and exact the payments at the end of our purchasing experience are introducing new functionalities and use cases and related risks (Espinoza 2020). As of today, we can easily pay with a credit card or with a smartphone, but also cashier-less supermarkets are already on the rise, with IoT cameras and machines that are monitoring what we put in our supermarket trolley and automatically settling the payment as we check out of the shopping center. The same is already available for watches and bracelets, for petty purchases on the go. Cyber risks could go from actual money stolen, to the capturing, in full transparency, of the data explaining our movements and purchases—anything from hacking to scamming and stealing of bank account details, social security numbers, and other sensitive personal information. If all these are cyber risks we run in an multiplicative way in a smart cities because of criminal organizations, other cyber risks could instead be related to perfectly legitimate ones, but with a fine line that separates a value added solution we may like (even if we may have never asked for that) and the influencing and manipulative consequences of the commercial use cases that third parties could be pursuing through the extended, real estate embedded IoT networks that are now becoming commonplace. Who could judge if this exchange was agreeable and let alone fair?

In fact, one of the obvious battlefields, in the public spaces of smart cities, will be at the shopping center and related experience level. In a digitally interconnected mall or store, we may walk and spend time unaware of our being monitored by facial-recognition cameras, Bluetooth trackers, smart sensors, and self-service tills—all interpolating cameras, smartphone data sucking devices, and else. On principle, the retailers, or the real estate investors owning the building and related space, could harness data and manage IoT networks and AI-integrated systems to improve our consumer experience and overall well-being (Fern 2020). But its final use and overall cyber security could also be contentious as shopping center and stores are becoming high-value targets for data thefts and unacceptable uses. From a cyber risk perspective, software vulnerabilities could allow attackers to gain additional access or escalate privileges, leveraging one of the many point of entry of an extended network of IoT devices. Among these, those running on NFC (near-field communication—a technology often used for contactless payment) and RFID (radio-frequency identification—often used for inventory management and tracking) could be more easily read and overwritten. Apart from thefts of money that could regard payment systems specifically, customer data are also a rich prize, the more valuable the greatest the personal detail of an individual, with very liquid (and efficient, albeit illegal) "black markets" that are used to monetize these for commercial use or the hacking and impersonation of victims. In order to limit this, the IoT systems which could be built inside-out the real estate building and sold as an integrated part of its attractive proposition should consider "privacy-by design" approaches, with encryption and firewalls, segregated data storage and offline backups, strengthened Internet and Wi-Fi security, and frequent updates and the use of penetration tests to check the inviolability of the system. Even more, they should cover seamlessly the digital and physical dimensions, with the security systems of offline places becoming integrated with the one of online channels and data bases. Eventually, even if immaterial, cyber risk should be treated and addressed as part of the safety standards of a home office, building, neighborhood, or entire city—and priced accordingly when valuing a real estate asset. And massive attacks should be prevented, timely predicted, and managed when they occur with the same dedication and gravity of earthquakes or other natural disasters. In fact, in a cy-phy city, a holistic risk management approach should cover consistently both the digital and physical dimensions and their main risk interrelations.

8.6 Towards a New Model for Cities

No matter what the technological development and digital innovation that were already available, cities across the globe have been hit hard by the coronavirus' death and disease waves and will be hit harder by climate change (Wilson 2020). In truth, as reminded by other authors, the history of cities is also an history of pandemics, pollution, and wars (in them the enemy tends to hit harder to inflict major social and economic pain—see the example of the V2 for London, and then the bombardments of Berlin during World War II). Still, for centuries it has been hard to stop single individuals and entire families leaving their bucolic calm and rushing to become citizens. As reported by the *Financial Times (The enduring attraction of cities* 2020*)*, the equivalent of eight cities the size of New York were being formed annually around the world before the COVID-19 outbreak. Because of COVID, clustering has now become a high-risk activity, with many a journalist asking if we are going to see, as already apparent from the real estate agency listings and most recent deal trends, a reversal of the centripetal momentum that has drawn people and business into urban clusters in the first instance. Indeed, the death of cities and urban doomsday has been prophesized many times before, but videoconferencing and augmented reality over the web makes it more plausible this time around. While the prediction (and fear, for collapsing property values, abandoned offices, and empty restaurants) is understandable, it ignores the vast gains in wages and productivity that were fostered historically by cities. As reported by the *Financial Times (Harding* 2020*)*, a doubling of city size raises productivity per head by 3–8 percent—and that implies that a city of 16 Mn inhabitants will produce 30 percent more pro capita than a city of half a million people. This increase in productivity has been investigated and can be traced down to the fact that it comes mostly from working "in person" in a bigger city, with more constructive relationships made possible by the short distance and also by the greater multitude of specialist services that offer a better matching of workers to jobs and companies to customers. Eventually, if new learning, innovation, and matching can only take place when urban life resumes, some change will however be needed.

Flight to cities has long been an engine of growth across the world and, more recently, even more so in the sprawling metropolis of the developing world. But COVID-19, which flourishes in areas of high human density, is now forcing governments in developed and emerging economies to rethink their approach to urban development (Wheatley 2020) and to the overall design and management of large cities. In the more developed countries, take France or Italy, for example, a new concept of a "15 minutes city" has got much in vogue (Whittle 2020). In short, as it happened after the 1570s, when a plague tore through Milan, everything had to change, with a large church, the lazzaretto, becoming a hospital, but not only. The same area (the "lazzaretto," a noun now associated in the Italian language with the very idea of a plague-infested area) is now at the center of an ambitious urban experiment, announced in April 2020 by the mayor of Milan, Giuseppe Sala, and confirming the lead of Milan as a rising city star across Europe and not only. The "lazzaretto" area would host a pilot scheme for "rethinking the rhythms" of the financial, industrial, luxury (and likely new, contemporary leisurely life—the "Milano da bere," as opposed to the "dolce vita" of Rome) capital of Italy. The pilot scheme will test the very idea of "rethinking the rhythms" of working and living in the city, offering services and quality of life within the space of 15 minutes' walk on foot from home—also at the light of creating a less polluting and more circular-regenerating ecosystem, where little time is wasted in commuting and almost zero on CO_2 polluting travelling. Work, home, entertainment, education, and healthcare are all meant to be available in less of a quarter of an hour—all based on a better urban organization and supported by enhanced connectivity and "augmented reality" digital communication and, where needed, regeneration programs, with little or no cars going around to reduce emissions and waste of time. If this is going to happen, the strength of Milan will reassert itself, and everybody will be able to enjoy more productive and also more balanced lives, with greater wealth and wellness.

In developing economies, the challenge posed to the transformation of cities and urban planning is even more evident. According to UN data, more than half of their population already live in cities, some of them chaotic and highly polluted and characterized by extreme wealth polarization—inequality instead of inclusiveness. More than a 15 minutes' walk cites, places like Nairobi or Mumbai, could be better defined, as of today, like a 15 hours' drive cities (the time required to do a few tens of miles during busy days). Sustainability is not just a matter of speediness, or efficiency—as we have discovered during the pandemics. In fact, according to the UN, at least 90% of COVID-19 cases have been recorded in urban areas—driven by their population density, but also by their sub-optimal standard of living and by how this density is (poorly) managed. In fact, as stated by Sameh Wahba, the built environment can transform density from overcrowding to livable, healthy, and attractive density. As a result, the pandemic has become a catalyst for cities from Bogotá and Buenos Aires to Chennai and Jakarta to rethink their approach to urbanization and to minimize the cities' vulnerability spots (Wheatley 2020) and, hopefully, increase the attractiveness of living there also during "normal" times. Most efforts, to turn these weaknesses into drivers of sustainability, are now rightly focusing on the benefits of environmental policies, green conversion plans, and smart city technologies, including an "open data" approach and the widespread use of AI. Action plays include new mobility infrastructure and cycle lanes, vertical farming and agriculture being brought back to urban areas, the reconstruction of entire buildings, and the development of better social housing as an alternative to suburban slums. This is a long journey indeed, and it will require synergies across finance, real estate, and technology, as well as a good degree of creativity and inventiveness.

8.7 The Leopard on Mars

In this book, we started our journey from a provocative critique of "smart cities." They were a bit trendy before the pandemics (for Italian cities with more than 15 thousands inhabitants, 36% of these had at least one smart city project in 2018 and 42% in 2019 (di Milano 2020)) and could become even more so after that. In truth, they still look mostly in their design and development stage and in continuous evolution even with regard to their extended conceptualization, with some best practices that are maybe more advanced but with a great majority of medium-sized and small cities that are not even aware of what the term means. In the land of Giuseppe Tomasi di Lampedusa's *The Leopard*, it's not obvious how many cities could reasonably aspire to be named as such in 5 or even 10 years' time. Not only the COVID-19 pandemic may have been creating the basis for an accelerated change in cyber-physical relationships, with people accepting more easily being monitored (but with them being in control and with clear benefits for all and on a fair-trade basis). But also, it has also been creating huge waves of poverty and great unemployment (if we include the furlough schemes), with brutal disruption in the normal wealth creation proccss of countries and companies. They will all be left clogged with unmanageable debts at the end of this health and economic crisis of (in the words of Mario Draghi, past President of the European Central Bank) "almost Biblical proportions." With so many zombie companies (and countries?) going around, the question is then "how to get this investments started, if the money is not even enough to keep the current set up going?". How can we think of a cyber-physical realm when connectivity is poor and social distancing is actually raising the data divide, even in the offline world? And, even assuming new funds will be available to plug the main gaps, to allow most households to access Amazon and Google and Facebook (not to mention the upcoming Alibaba, Tencent, and Baidu), how could we make sure that this "smart" development is not really just as such in the eyes and for the interests of these, already super-rich and potentially oligopolistic digital platform powerhouses, able to cut across geographical borders and technological standards, the different rules of law, and even geopolitical spheres of influence? Are these, hoped for, more and more achievable changes all aiming, in the land of *The Leopard*, to just keep changing everything so that nothing does at the end? That has been a great part of the historical resilience of cities and citizens that are hard to change, for the good as much as the worst.

Still, the pandemic crisis and, among other factors, the deployment of the New Generations sizeable fund made available by the European Union make it an almost unique opportunity to change this and accelerate the transformation required—to really change something in order to disrupt everything. Indeed, the move towards a smart city, and then cyber and meta, and ultimately "sustainable" in our "cyber-physical" and most extensive definition, is a long one and by no means assured. Tragic events such as a regional war can derail that project and send us back (or forward) towards an age characterized by our short awakening before a second sleep. Potentially terminal events, such as climate change, could short-stop our fancy vision of wealth and well-being creation and lead to a reinterpretation of "sustainability" in a more restrictive way, as a humankind preservation plan in caves underground or as a long-term project to colonize Mars (with one way Space-X ticket and driving a fitting Tesla Cyber Truck over there). Given such challenges and level of uncertainties ahead is maybe worth now, at the end of this book, to recap what things could and should happen to support that journey toward our envisioned "cy-phy" sustainable city. In short, centralized BDP (big data platforms) have already been touted as the key and most powerful component of "smart cities," with a mix of centralized, decentralized (or off-centralized—e.g., referring to the opportunity to centralize data and applied analytics over the cloud) approaches that is likely to emerge. Specifically, decentralization may refer to the emergence of multiple data and intelligence centers or even just focusing on some of the applied analytics component. A "smart" city, we have argued, could become more cybernetic (hence "cyber") if aimed at capturing and receiving a greater amount of feedbacks from the physical environment—to establish a two-way level of communication between the world of atoms and the one of bites. Also, we have argued, the city would become "meta" if this could aspire to provide not just "horizontal utilities," good for most citizens but for basic functions such as mobility, safety, or energy consumption, but could also articulate solutions aimed at the specific citizens and built around her and his personal needs.

Take, for example, traffic in the city, as an obviously relevant starting point to address. As also noted by Anna Kong Jerlmyr, Mayor of Stockholm (Jerlmyr 2020), flattening the peaks in demand for urban infrastructure, with the help of data and IoT networks capturing them in real time, could help in making the city more livable, valuable, and sustainable. This would not just happen because of the information available on traffic (which has been available for a while) as this will need to be coupled with other "smart-tech" solutions, e.g., planning the return to a mixed way of working and studying (partly with physical presence, partly remotely and with digital links) on the basis of this information and offering the right kind of incentive to align to the plan. In Singapore, for example, drivers pay automatic congestion tolls based on traffic with "electronic road pricing." Similarly, discounts or free coupons could be offered for smart working or for travelling on off-peak times or secondary roads. This simple example could then be extended to many others, such as healthcare, leisure, and holidaying and much else, to progressively address the issues of the physical real estate and infrastructure of the cities to then progressively shift to the needs of the individual people. One thing is to plan and use data and AI to optimize and flatten the usage of a bridge and shave off the peaks on energy consumption. But another thing altogether would be to advice and use data and AI to optimize my health and well-being, leisure, and attention at work or in education, based on my specific needs and on how I can leverage all the resources, physical and digital, made of things and of other people the city can offer me. Advice could be offered on the way I plan my movements and my activities, eat and work and meet people and rest, choosing how to wear and how to communicate, selecting a few options across the spectrum of physical and digital tools to connect with others.

In a more compelling vision, this increasingly cyber-physical and meta pursuits would then aim to make the city (as the proverbial microcosm where society and markets develop and where most of humankind is going to live anyway, given its secular urbanization process) more "sustainable." In our definition, sustainability would address targets of wealth, well-being, but also fairness and inclusion, in a hosting environment that is meant to protect us as we preserve the hearth for our future generations to come. In this transformation journey, key drivers are certainly finance, real estate, and technology, and there is a virtuous circle to be considered, to support an optimal path towards wealth-being. More importantly, and in an accelerated way after the COVID-19 crisis, the development of "open data" approaches and AI-based solutions, able to interconnect to full fruition the digital and physical world, are becoming even more important. No matter what the poverty left by the pandemic, places that were left behind should just aim to leap-frog the different stages discussed and get to the target end state in time. Palermo and Sicily, the land of *The Leopard*, in our example, with a 5 Mn population, a number of physical tangible and intangible (e.g., arts, history) assets, and a very cheap starting point in the investment cycle, could ideally be set as a "special economic zone" or "sandbox" to design, jump start, and test the world of the cy-phy companies and cities to come. With access to "open data" pools granted by data owners and with certain rules to treat data in full compliance with GDPR styled principles, AI-based apps could be built to address the most critical and relevant hard problems for Sicilians and for the rest of the Italian population. Cy-phy solutions for hard problems are not dissimilar across the globe—hence exportable.

8.8 A Bit of Clouds Could Help

The development of an "open data" framework will require a proper regulatory framework, standardized technological protocols and API, and—as discussed—some kind of economic modeling to provide indication of what a "fair" price for data and for their different uses could be, at least in an initial phase and until an efficient, liquid, and transparent market for data and info-telligence has been developed. This "open data" framework would then mark a major disruption in most industrial sectors and likely drive the new value migrations to come and the emergence of the cy-phy winning trillion-dollar companies of the future. It will be a transformation process at scale, but not necessarily limited to "at scale companies." In fact, a bit of "clouds" (as public or private or as multi-cloud mix of solutions on offer) could help. A cloud-based approach to "open data" would more naturally support this—as the hardware and software are also open in their access and infinitely flexible as already at scale when virtually pooled and set as a service. Such an approach would take away the scale conundrum for small and mid-size players of all kinds, allowing them to focus on their mission and perform better. They should also allow super safe and global standards, for owning and treating the data, to emerge. They would support the specialization of the most capital-intensive components running across the info-telligence value chain. This, in turn, would support the vision of companies which are virtualizing a number of critical capabilities that are stored, developed, and managed through the cloud. They would retain their brand equity and invest in their distinctive value propositions, be it their last mile, physical relationship capabilities with end customers, or their ability to design better use cases that can be delivered mostly online and in real time through 5G.

Because of this migration towards the cloud, the digital transformation of traditional companies into cy-phy ones could even happen more effectively and efficiently. In fact, digital transformation is a bit of a misnomer, and, leaving aside the hype promoted by most management books, not much is happening in real life yet. Changing a traditional organization with its legacy systems, internal processes, culture, and internal bureaucracy may look sometimes like draining and refilling a lake using a glass or a carafe. Something similar could also be said for cities, with a lot of their getting "smart" statements happening just on paper, given the multiple constituencies and technological legacies, not to mention the different agenda of ever-changing policy makers. An "open data" approach would instead facilitate the engagement of governments and businesses, people, and civil society at large. If the much hyped digital transformation is the way to lead the change associated with the application of digital technology in all aspects of society, then a strategic and operational shift to the cloud and with an "open data" approach could stand as the real discontinuity allowing this to happen, using powerful hydro-pumps instead of a carafe. In this context, cloud adoption is then, off course, just a mean to an end, with business drivers such as exceptional user experience, accelerated time to market, higher service quality, cost leverage and scale, repeatability, flexibility and safety, security, and compliance with regulation mentioned as the most frequent winners and justifying this kind of strategic decision (Abdula et al. 2018). In fact, cloud computing is, according to the 2011 National Institute of Standards and Technology (NIST), a model for enabling ubiquitous, convenient, on-demand network access to a shared pool of configurable computing resources (e.g., networks, servers, storage, applications, and services) that can be rapidly provisioned and released with minimal management effort or service provider interaction. On top of this, cloud computing has become an enabler of business and operating model transformation that could be applied to "sustainable cities" as well—and even more so if they move towards "open data" approaches.

In the move towards building a truly sustainable city, leveraging cloud computing capabilities on the basis of an "open data framework," collaboration and co-creation with citizens will become even more crucial to the success of the place-location and unique value proposition of the city itself. It could also accelerate innovative disruptions, across all industries and sectors, and to unprecedented levels. An early example of this "open data" approach is decentralized finance, also called "open finance," a better term by all means. Open data, in its extensive definition, is a blanket term for financial services like deposit taking, lending, global finance, and asset management that are designed and run leveraging decentralized technologies (mostly resident over the cloud) and smart contracts (mostly using decentralized technologies like block chains—where contracts are made of computer code). The innovative disruption of open finance would allow things like the issuance of real-time loans, with real-time (or pre-approved) analysis of the risks underwritten, leveraging data on customers behaviors—captured online and mostly offline and in the context of cy-phy cities. This loan offering and related investments could come with no middleman in between and little influence left for central bankers on the overall money supply. If new cryptocurrencies are created and used to price a trading opportunity, then their earned return would be directly related to the investment they take up and bypassing most supervisory activities and the monetary policy making of central banks. Companies and SPV managing large projects (such us a city infrastructure or a building development) would then be able to issue their own shares and bonds and any other ad hoc hybrid structure, without having to deal with financial intermediaries, not to mention lawyers (Won 2019).

Open finance would also lower most physical risks—reducing the chances of robberies from banks' branches, for example, as money is fully digitized and traceable. In a sustainable city setting, it would represent a further step in the digitization of financial services and of all kind of transactions, and it could also reap the benefits of scale, across cities, if the cloud solution on offer is global. Or improve the level of service on offer at city level, if a 5G technology is developed around a nearby hub (single cloud or multi-cloud) to ensure almost real-time and an almost disappearing latency. However, it would also, for financial services and for entire cities, broaden the potential attack surface for cyber risk-related negative events that can lead to thefts of money and loss of consumers' confidence (even more so when, on top of money, personal, confidential information is stolen). Rapid adoption of cloud and open data by financial service and other companies, and even entire cities, would then imply a further shift of power and relevance towards the bid data platforms and computing applications managed by technology companies, with all its good and bad. This further polarization of digital resources and capabilities are also posing a significant concentration risk, if all this is, in the western world, concentrated across three biggest cloud providers—Amazon Web Services, Microsoft Azure, and Google Cloud (Noonan 2020). These players have certainly the largest scale, and likely the highest standards of security, and they can develop more that can be shared across all kind of players, e.g., more sophisticated authentication features including facial and voice recognition, geolocation, and even tools that evaluate how a phone is held and how keys are pressed to determine by whom. But would this significant concentration make cloud, open data, and the city itself more sustainable? Or just build the premises for the next catastrophe to come or oligopoly of the few to bear?

8.9 Defining Moment

As the world is facing a second and maybe third and fourth wave of the COVID-19 pandemic and looking forward to the mass implementation of a global vaccine, we have been wondering if this is the end of the urbanization trend we were all relying upon. And if the death of urban city centers is something we should now consider, with digital technology acting as a good substitute of the coronavirus vaccine. Among city centers, a *Financial Times* research has also shown how the truly global metropolis—such as New York and London—have suffered the most (Romei and Murdoch 2020): with empty shops, cafés struggling for survival, and entire financial centers looking like ghost towns. Few cities have in fact escaped the impact, with visits to central Paris down by 40% by mid-October and even Stockholm, mostly unconstrained by regulatory restrictions, also down by 20%. Overall, the footfall declines at cafés, restaurants, retail, and well-being and leisure venue have been brutal across the globe, with some rebound in the summer, followed by another decrease in the fall. Certainly, main global cities are reacting and trying to reinvent themselves—it would be a gross mistake to write off New York and London, with their long tradition of changing shape to get even more competitive.

The City of London, for example, is trying to encourage small businesses and the art sector to get back, in new and more innovative ways, more in tune with the times—promoting start-up hubs and more affordable and safer workplaces (Thomas 2020). As part of this, a strong indication has been given on having buildings better designed for remote working practices, with spaces to socialize and meet, with better pedestrianized and bike routes, to shave off picks in public transports, car congestion, and pollution—but also to make the switch between smart and office working more dynamic, immediate, and easy. More ambitious discussions are then promoted among a vast array of stakeholders to rethink technology and regulation and fiscal incentive and more towards reimagining London for the next century. Before the pandemics, the City of London was facing a scenario of polarized and unbalance (and unstainable?) future, with ever-spiraling property prices, served by a "disenfranchised population" painfully commuting-in daily from elsewhere (Editorial Board 2020). After the pandemics (and with COVID-19 still running), the city is a near-ghost town, with bankers and lawyers working from home offices and suburban areas (or the countryside), not to mention the approaching Brexit. To make up for this, a new vibrant vision needs to be agreed and pursued by many stakeholders. Leaving aside the regulatory setup, and the fiscal incentives to innovate which are not part of our analysis, the main challenge will regard reinventing the physical and built environment, given the uncertain future of a truly post COVID-19 world. Even if a vaccine is developed in the short run, the future of many professional workers is likely to be a mixture of working from home, with a few days spent in the office each week. This would allow companies to plan dynamically the usage of office space and shrink this overall accordingly, lowering rents in absolute but also percentage terms.

This is no small challenge, as it will entail persuading workers to commute into town from suburban areas, with more attractive offices and streets where to spend time—all with flexible use in mind and, potentially, with some repurposing of offices into residential buildings ready for short stays of 2–3 days, and with a bit of amenities open in the evening to allow a more proper social life, when staying over for a short "work-at-the-office" trip. The lack of details of what future working life will be and how it could all work is making this planning more difficult. In fact, while views on the future of big cities, as winning agglomerate, so far, of skilled workforce and disposable income, vary greatly, it is becoming a safe bet to assume some great change ahead. The pandemic and even climate change and the inherent fragility of big cities vis-à-vis any other major negative event (wars included) are all pointing to great disruption to come, reshaping cities and urban and suburban areas. Whatever the final endgame, with shifts to more remote locations, wider residential spaces and a growing "work from home" share of the population, it is clear that digital technology, as embedded in infrastructure and real estate and as a core tenant of urban and suburban regeneration and development plans, will be a key driver of future success. Both should enable more space for innovation and flexibility in the way we communicate, meet (virtually or physically), and interact. Many urban centers are already getting ready for this big change and to get "smarter." Paris is championing the concept of the "15-minute city," while Milan is actually proof testing it. Melbourne is proposing a "20 minutes neighborhoods"—that is all about "living locally," giving people the ability to meet most of their daily needs within a 20-minute walk from home, with safe cycling and local transport options, while Montreal is working in the meantime on defining a hybrid system that combines remote working and the continued use of physical space, with cycle lanes warming up in winter and with shield to protect cyclists from ice and wind. The overall expectation is that this "getting smarter and more digital" will eventually lead to downsized and smaller, more flexible workplaces (with conference rooms, hot desks, and even moving walls and panels) and larger residential areas, with greater and better shared facilities (including "green ones," such as horizontal gardens but also vertical groves and farming—we also discussed the idea of integrated buildings).

Indeed, we believe that the coming years will be a defining moment for cities to think, plan, and act for their substantial redesign, from downtown to suburban areas and encompassing all dimensions, from bricks to cables, from atoms to bits. Indeed, this challenge of change (because of and following the COVID-19 pandemic, a defining moment for many years to come) is a unique opportunity to make our cities, their center and larger urban and suburban areas, greener and healthier, more sustainable, and attractive from a wealth and well-being point of view. We should have not waited for COVID-19 to have additional bike lanes, better hygiene on public transport, and faster adaptation of low-emission engines. We should not have waited for the pandemic for redesigning office and commercial and residential spaces to make them more adaptable and flexible, working around the individuals that are going to live them and not fitting them into what has been built—we all have fit at the last minute and quite imperfectly our office properties as part of our smart-working from home, but we have now the chance to redesign our home as also partly "smart-office"; and we may have to redesign our closed down offices into something more residential and as "living and leisure" center, if we want to go back to the office for what it really brings as value added, e.g., the opportunity to interact with other people and work closely with our team mates (and accidentally provide chances casual encounters, for socializing and romance or serendipity opportunities). Hence, the rigid separation lines between offices and homes could become blurred, as well as those between the city center (and the many city centers of the 15-minute walks cities) and the larger urban and many suburban areas: everything will get multi polar and redistributed, leading to a new ways of designing, building healthier and more resilient cities that should ideally lead to greater, sustainable well-being. The pandemic crisis would not be wasted if this could be an endgame of it.

8.10 Cities as the "Cy-Phy" Platforms of the Future

A new grand bargain in the hi-tech digital sphere could soon be at stake, according to *The Economist*, as a consequence of the accelerated decoupling and confrontation between the United States and China's political, economic (and military?) forces. In fact, in the recent past, the battle for geopolitical dominance has been fought by the United States and China as leadership in hardware and software and in their respective abilities to integrate them all in a "system of systems" (The new grand bargain 2020). In business (the focus of this discussion), the United States has two "big digital" companies worth more than 2 Trn and three more worth more than 1. But China has a population of 1.4 Bn (and little interest in data privacy), giving it a supremacy on data pools and AI, with tech giants as Alibaba and Tencent already covering most of the main, traditional industrial sectors and going beyond that. It is also moving aggressively on quantum computing and 5G, a new generation of mobile networks where Huawei is said to have gained a clear hedge and maybe some technological supremacy, but also (this is the worry of western governments) potentially acting for the comfort and support of China's autocracies. Given the increasing relevance of the digital dimension, even new geopolitical equilibria are now being discussed in terms of digital trade zones and relationships. This was just accelerated by COVID-19, which has empathized the role of digital alliances, by pushing much of human activity into the cloud, and with the ensuing (obvious) protectionism driving towards a "splinternet." While our geopolitical view of the world BC was still mostly reliant on geographical aspects of the physical realm, including natural blocks such as mountains, rivers, seas, and lakes, in an AC world blocks can be better interpreted as technological standards, digital firewalls, and cybersecurity defense (or attack) systems. If BC geography was our main destiny, AC the digital ecosystems will be as much.

Digital platforms are then taking center stage, also driven by their own politics and governance, with some more open than others and attaining (in the case of big digitals like Amazon, Google, Microsoft, Facebook, or Apple), an accumulated wealth and related tangible and intangible power that is bigger than the ones of most countries. In fact, even countries could be now better understood, as they become more and more digitized and "smart" and interpreted as platforms—as a set of national operating systems in accordance with a local law that is defining common standards and acceptable end uses. Even the wealth of nations analyzed by Adam Smith are becoming more and more driven by their digital resourcing, from skilled people to computing power, to big data sets available for AI mining to connectivity bandwidth and velocity. All traditional industrial sectors are also becoming more digital and actually blurring their traditional boundaries in an "al-out" competitive fight, from transportation (competing with communication) to healthcare (competing with wellness), from teaching, politics, and welfare where robots, androids, and the like are increasingly substituting human resources, let alone influencing their key decision-makers. Given such a polarizing scenario of extreme digitalization, it would be too easy to underestimate the potential role cities will have in the future. In fact, there is a great future for them if they will be able to reinterpret and reinvent themselves as "platforms," but on a physical and digital basis. This is the critical hypothesis for discussion of this book. If offline and online information can be captured, stored, shared, and analyzed while running at speed on OpenRAN mobile networks (that allows networks to operate on mixed and matched multiple carriers, avoiding the systemic dominance of a single player a là Huawei), then cities can have a bright future. In a global competition between business platforms, Google could be juxtaposed with Milan and Amazon with Berlin. They could coexist successfully, but without allowing the dominance of digital-only players. Competition would then be driven (again) by the emergence and prevalence of great cities that can sustain the wealth creation of their businesses and optimize the wellness of their citizens—in a "cy-phy" way.

8.11 Conclusion: From Smart to Sustainable Cities

For all the talk of "smart cities," as technology platforms capturing data and information that are meant to support the growth of societal civilization and the wealth of the local economies, they have so far remarkably failed to deliver on these ambitions. Not only the basic idea of "smart city" seems to rely on the employment of digital technology for the narrow offering of public and private services that are used to optimize mobility, safety, energy consumption, and few other utilities, with limited scope for delivering more relevant and critical solutions to its citizens. But also, the few developments achieved so far seem to put little emphasis on building resilience in the city—not aiming at ensuring its sustainability in a world increasingly dominated by extreme risks (e.g., pandemics) and existential threats (e.g., climate change). In fact, based on a number of discussions with players of the finance, real estate, and the telecom sectors and with civil servants of the central and local public administration, the very concept of "smart city" looks already "passé" and in need of some rethinking—starting from its naming and labeling. First, cyber (and cybernetics) looks a better term to describe how new digital technologies could be used to design and manage the interconnected digital twin cities, to optimize the ability of the physical one to create new wealth and well-being and increase its social and economic resilience during challenging times—as it has become all too obvious during the COVID-19 crisis. Second, the framework of discussion is also suggesting developing the concept of "cybernetic city" into a "meta" one—meaning, not only digital systems that can allow physical ones (e.g., cities and citizens, households and companies) to become "smarter and smarter" but also with physical ones being able to optimize their digital dimension, as offline data and real-life behaviors can support the optimization and development of the cities' virtual brains (as a sum of "open" big data and artificial intelligence apps all sitting "over the cloud").

The sum of all this is a "sustainable city"—a city where the digital and physical dimensions gets more and more enmeshed together, seamlessly and continuously to increase resilience. In the cyber-physical setting of "sustainable cities," our experiences, and our behaviors as households and companies, become inextricably defined by a mix of bites and atoms. People compete in their working environment and social life because of their prowess in mingling and optimizing the two "cy-phy" dimensions together. And the winning companies are also emerging as better apt at leveraging the two dimensions together, in order to provide superior value to end customers, mixing on- and offline data to provide the best information-intelligence (info-telligence) available, to solve hard problems and provide serviceable truths. In turn, such "cy-phy" sustainable cities should be able not just to perform better during normal times but also to navigate better through crisis time—as we face a number of existential threats that need to be addressed and solved—starting at the level of the cities, given the urban population living in urban areas projected to get to a 75% at global level. A clear understanding of the fundamental dynamics of this emerging "info-telligence" economy and of the potential dominant role of "big digital" players is then required to discuss which kind of valuable use cases could be designed and built, within the sustainable city, around individuals and to improve the success and the resilience of its households and companies. Such use cases could range from sustainable healthcare to sustainable finance, to sustainable real estate, and to sustainable data monetization business models that could be identified across all kind of economic sectors and dimensions of social life. On this basis, a number of winning companies (not necessarily the current "big digital" players of the western and eastern world only) could emerge to create huge value in the market and for their stakeholders, by leveraging the principles of the info-telligence economy in the context of their targeted needs. These rising "cy-phy" companies could then become the next trillion-dollar companies as competitive relevance is shifting from brokerage and connectivity to data and then AI-based capabilities that can be put at use to address critical problems. This was the vision before the pandemic and it is still today, after the COVID-19 global outbreak, but with a dramatic acceleration in place. As we all strive to reach the "other normal(s)" after the pandemic and given the climate change challenge, we all need, now more than ever, to become more "cy-phy" and leveraging AI to its maturity. This has never been truer for cities as well, as they will need to compete not just in a recessionary and globalization-disrupted environment, but also to fight for their survival as epicenter of the humankind civilization and as business and social platforms. This is not a small challenge, as their transformation success is by no means assured and the chance of waking up in between a first and second sleep is a very clear and present danger.

References

The great disrupter, The Economist, September 18th, 2020.

Power in the 21st century, The Economist, September 18th, 2020.

Petrostate versus electro-state, The Economist, September 18th, 2020.

Free the data serfs, The Economist, October 26th, 2020.

Tear down this wall, The Economist, April 25th, 2020.

Finance and real estate wealth-being, BUP – EGEA, 2020.

Siddharth Venkataramakrishnan, Cyber risks take the fun out of connected toys, 25th November, 2020.

Adam Green, Smart homes revolution tests legal liability regimes, Financial Times, 25th November, 2020.

Internet of Energy powers up hackers' threat to electricity grids, 25th November 2020.

Javier Espinoza, Digital payments deepen the threat of online fraud in Covid era, Financial Times, 25th November 2020.

Nicolas Fern, Retailers face tough sell over data collection technology, Financial Times, 25th November 2020.

Ben Wilson, Metropolis, Penguin Random House, 2020.

The enduring attraction of cities, September 29th, 2020.

Robin Harding, Cities are too resilient to be killed by COVID, Financial Times, September 30th, 2020.

Jonathan Wheatley, Virus prompts rethink of developing economies' urban planning, Financial Times, September 22nd, 2020.

Natalie Whittle, Welcome to the 15-minute city, Financial Times, July 17th, 2020.

Politecnico di Milano, Osservatorio Internet of Things, February 2020.

Anna Kong Jerlmyr, "Flatten the curve" of traffic to make cities livable, Financial Times, October 19th, 2020.

Moe Abdula, Ingo Averdunk, Roland Barcia, Kyle Brown, The cloud adoption playbook, Ndu Emuchay, Wiley 2018.

Daniel Won, The definitive guide to DeFi (Decentralize Finance), December 22nd, 2019, Exodus.io.

Laura Noonan, Advancing bank technology "broadens hack attack surface", Financial Times, March 22nd, 2020.

Valentina Romei and John Burn Murdoch, From peak city to ghost town: the urban centers hit hardest by COVID 19, October 15th, 2020.

Daniel Thomas, City of London seeks to "reinvent itself" after pandemic, Financial Times, October 18th, 2020.

Editorial Board, A vibrant vision of the City of London's future, Financial Times, October 21st, 2020.

The new grand bargain, The Economist, November 20th, 2020.

Chapter 9
Finance, Real Estate, and the "Smart City" Challenge in Milan Area: Toward a Sustainable Future?

This viewpoint is based on the round table discussion held in Milan on the 12th of February 2020 at Cracco's Restaurant. Chatam rules' applying, this short annex reflects the summary of the discussion, with over 35 executives joining from Finance, Real Estate, and Technology.

9.1 Finance Role in the City's Development

The Real Estate sector, despite being one of the oldest businesses in the world, has seen little innovation in the last 50 years. Building techniques as well as business models and financing methods remained generally unchanged during years of exponential and drastic developments in other sectors (most notable transportation, manufacturing, and information systems). It is for that reason that in this sector resides a great opportunity for innovation and, being real estate a constant presence in the private and public life of every human, an opportunity for re-thinking work–life balances and daily routines.

Banks play a key role in the real estate industry, being exposed to real estate assets in two main ways: first through the possession of proprietary assets that, due to the increasing digitalization of the sector, need to be enhanced and converted to more useful destinations of use (Italy's top 10 banks keep almost €20 Bn of Real Estate assets such as offices, headquarters, and local branches in their balance sheet) and secondly through the credit exposure of mortgages derived from their core lending activity.

It follows naturally that one of the main objectives for banks nowadays is to internally develop a more and more peculiar ability to evaluate Real Estate assets and development projects, through a smart use of available data but also through the adoption of a real-estate-alike attitude and sensitivity. This objective certainly requires an internal cost for banks in order to build a new area of expertise, but the amount of internal resources allocated in real estate assets is so significant that this

© The Author(s), under exclusive license to Springer Nature Switzerland AG 2021

C. Scardovi, *Sustainable Cities*, https://doi.org/10.1007/978-3-030-68438-9_9

topic already attired the attention of top positions inside those organizations, since it will enable them to better understand the true value of development initiatives and thus steer capitals in the direction of project quality and profitability.

An example of the proactivity in the direction of optimizing real estate assets inside banks can be found in Intesa Sanpaolo. Being the biggest bank in Italy, with total assets worth more than 850 billion euro (June 2020), Intesa Sanpaolo found itself in the position of being involved with two of the major real estate development projects in Milan: MlanoSesto and Milano Santa Giulia. Despite the difficult situation, the exposition of the bank led to the decision of pursuing the investments, partnering up with two main real estate development companies in Italy: Lendlease for Milano Santa Giulia and Prelios for MilanoSesto.

9.2 Big Data's Role in Real Estate

The importance of data in today's world is clear to anyone dealing with business decisions: data permits to have a "scientific" approach to the analysis of a particular topic and to make rational choices that create value for the company and its stakeholders. The use of data is a hot topic in the real estate industry, since it allows for a better evaluation of assets, a competence that is key to institutions like banks and investment funds, that keep a great portion of their assets in physical buildings. A clear evidence of this "hype" has to be found in the increasing relevance of the so-called "proptech" (property technology) sector, in other words, the application of information technology to the real estate market, in order to improve data collection, transactions, and legal compliance.

For what it concerns banks, in particular, having a way to automate and improve the quality of the valuation of real estate assets becomes key in order to be compliant with regulation's capital requirements and in order to keep updated the value of nonperforming loans (NPL) kept in the balance sheets. The process of evaluating assets, being it repeated constantly in time, represents a significant cost for banks, that are working, with their risk management department, to automate this process with algorithms and software.

At the same time, banks, as well as new actors in the real estate industry, such as crowdfunding platforms and instant buyer funds, are interested in the use of models to better predict the price of assets before investing in it (investing as collateral in the case of banks) in order to assure their business model will be economically sustainable.

What becomes crystal clear to the eyes of risk management professionals and investment funds' managers is the importance of the quality of data, since a good evaluation model needs to consider not only "standard" indicators such as the price per square meter of the area, but also atypical factors, i.e., the energy efficiency class of the building, its age, the potential for expansion of the neighborhood and local environment data for a long-term sustainability of the asset. This will allow their institutions to better understand trends, manage real estate developments, and allocate/divest funds on an informed basis and with a forward-looking perspective.

Another important use of data, in the real estate industry, concerns its adoption in the analysis of behavioral trends: studying the way people use buildings, architects first and developers then, can adapt their constructions to improve the services offered to the standard user. This new possibility will radically change, in the next future, the paradigms of the real estate industry, switching it from a product-centered to a service/customer-centered industry.

9.3 Real Estate Transformation: From a Product-Centered to a Customer-Centered Industry

In the most developed economies, in this moment, the trend of using data is becoming more and more common not only to calculate real estate value but also to extract information on the use of buildings, representing a clear transition from a product-centered view of the real estate industry to a customer-centered one, that means and requires to have the human being at the center of the system, thus developing buildings and related services with the objective of optimizing her/his well-being.

The old-school approach to real estate, already analyzed in the first paragraph, is also visible in the way in which, at the moment, most of the products (buildings) are often not suitable to the many evolving needs of its customers (people). The way in which people use residential buildings, offices, and other structures can change over time (let us think for example to the current COVID19 pandemic that is moving the working life of individuals to their private houses), and this represents at the same time a risk for the real estate investors but also an opportunity for new smart-city projects to adapt to this new paradigm, integrating systems that monitor the use of buildings and adopting building choices and construction techniques that favor changes in the designated use over the life-cycle of the product.

An example of the use of data to support building choices is evident in the CityLife project in Milan: before moving Generali inside the new headquarter in one of the three office buildings of CityLife, over 37,000 observations have been collected in order to understand the behavioral habits of the employees in the previous office. The data showed a clear evidence: employees spent more time in collaborative spaces like meeting rooms than sitting at their personal desks.

Studies and supporting data on behavioral uses of buildings like in the previous example can bring valuable insights and represent a fundamental factor of efficiency, cost reduction, and risk management in the developing phase of the building, allowing for smarter choices that meet the needs of the users and create a higher value for every stakeholder involved in the process.

9.4 The Sustainability Paradigm

The shift from a product-centered Real Estate approach to a customer-centered one is also the best way to meet the sustainability paradigm of the twenty-first century.

In fact, despite being largely discussed and despite having a large formal commitment from institutions of different types and sizes, sustainability is still far from being achieved, in particular in the Real Estate Sector.

Narrowing our view on Italy, in the last 20 years emissions generated by vehicles constantly decreased from 100,000 to 50,000 tons, while buildings' emissions increased from 50,000 to 250,000 tons. As those data show, sustainability inside cities, despite the recent years' trend of energy efficiency in buildings' development, is still far from being reached, real estate being one of the asset classes that mostly contributes to air pollution.

The sustainability issue generated by real estate also has dramatic impacts on the population's well-being: large cities like Milan become attractive due to their pace of change, their opportunities, and their innovation mindset, also derived from real estate developments, but building's emissions create at the same time an unhealthy environment with dramatic impacts on inhabitants.

Sustainability is a required collective effort that should start to be considered as key to protect each stakeholder's interest: health concerns derived by sustainability's issues may destroy a city's reputation, threatening therefore each actor in the value chain, from the final user to the developer, from the investor to the bank institutions. In fact, from a financial system's point of view, financing sustainable real estate developments is the only option to maintain a solid perspective in the long term, since this type of asset is going to remain on the market longer and with better performances in terms of efficiency and designated use, decreasing the risk associated with the financial operation.

Global warming, in the same way, is another long-term trend that creates possible risks for buildings and for cities in general and that should be considered as an additional factor to include in real estate-related due diligences.

Therefore, evaluating not only the financial risk of an operation but also its impact in terms of sustainability will be the new paradigm in the financial sector, a sector that should exploit the opportunity to show that sustainability is not a trend, but the only way of doing business looking forward.

9.5 Milan and the Smart-City Challenge

Overview

From being considered, just 15–20 years ago, as a "declining city, ugly and not attractive" from the "Rome" elite, Milan is now the most advanced and European city in Italy. With good working opportunities, an established network of notable public and private universities, a good presence of multinationals and financial firms, and being the European epicenter for fashion and design, Milan attracts young students and workers from the whole country and from abroad.

Milan's transformation in what we observe today is not just a matter of job opportunities and higher average salaries: in fact, those two factors were already present when the city was considered a declining one. What really contributed to the metamorphosis of the city has to be mainly identified in its urban restyling, deeply correlated with the presence of a broad and demanding mid-class, more actively involved, compared to Rome, in the municipality's decision regarding the urban landscape.

In the last 20 years, indeed, Milan observed various urban developments that helped to change its landscape to a more attractive and European one: just to mention some of them, the Darsena, an artificial connection of two open-air canals, better known as Navigli, gave the city a water-presence that was missing, compared to other European large-cities, and Porta Nuova, the recently finished real estate development that created a skyscraper-area inside the city, contributed to its luxury reputation and its international standing.

Opportunities

From an urbanistic point of view, Milan is a concentric city: its city center is delimited by the historical walls, then, the ring road, together with the mostly unused railway line that surrounds the city, delimits a second circle already urbanized from the metropolitan area of the city, that presents a lower density of buildings and that is expanding toward including most of the surrounding small municipalities.

An interesting and crucial opportunity, looking at Milan from an urbanistic perspective, is represented by the presence of abandoned railway stations and freight terminals situated along the railway line that divides the inner city with its periphery. In the nineteenth century, Milan had a high relevancy in Italy and in Europe in terms of heavy industry: in 1912, the municipality started an ambitious urbanistic project which comprised the construction of seven freight terminals around the city's periphery in order to better serve the local industries with an efficient transportation system. The project attended the expectations and increased the city's relevancy, but only for a limited amount of time: in fact, starting from the 1970s, the decreasing importance of heavy industry for Europe, together with the tendency of adopting road transportation methods instead of rail ones to move goods posed an end to the adoption of the railway network.

Seven of the rail yards once serving the city-surrounding railway system, namely "Farini," "Porta Romana," "Porta Genova," "Lambrate," "Rogoredo," "Greco," and "San Cristoforo," with an overall area of 1,250,000 square meters, represent nowadays an incredible opportunity for further city's developments due to their huge dimensions in hyper-strategic location at the border between the center and the surrounding area and due to their common public ownership.

Another opportunity for the city is represented by the reopening of the Navigli, a network of canals that crosses Milan and its province. Being not used anymore as a way to move people and goods from and to the city, due to the rising popularity of trains, at the end of the 1800 the city's canals have been covered by the municipality.

Today, just three of the canals, the "Naviglio Grande" and the "Naviglio Pavese," connected through the "Darsena" in the south-west part of the city, together with the "Naviglio Martesana," in the North-East part of it, are open-air and visible from the public.

In order to continue the city's transition toward a greener and more naturalistic one, the reopening of more canals inside Milan is seen from the citizens (with a 94% of positive feedbacks in a local referendum in 2011) as a good move to acquire a better-looking standing among the European capitals, with a water-presence distributed more densely among the different areas of the city.

Challenges and Recommendations

The many urban developments of the last years, the good public services, such as transportation, that the city offer, the multicultural mentality and the exploitation of opportunities (like the 2015's EXPO), helped to shape the city of Milan that we see today.

The urban development's opportunities identified for the future also contribute to a favorable outlook for Milan, which configures itself as the most promising Italian city.

However, with these opportunities come, at the same time, urbanistic challenges that need to be addressed.

The first challenge for the city is related to sustainability: as seen before, the contribution of buildings to air pollution has increased in the last 20 years; therefore, creating a place that, despite being attractive for the many job and entertainment opportunities, is not attractive from a health point of view. The municipality should address this problem with incentive policies for sustainable buildings: for example, by taxing less the buildings that create less emissions and by establishing collaborations between public utility companies and new smart-city developments, in order to create new areas that act as "champions" for the future of the city, with the delivery of eco-friendly services in the area of energy creation and distribution and waste management.

The second challenge, for Milan, is to keep the city open to residents of the middle and low class: in fact, many factors contribute to a high level of renting prices in the city, which most of the time are not sustainable by young people and families. Three of the factors contributing to this phenomenon are: the Italian tendency of buying houses instead of renting, the absence of medium-level real estate developments (just as an example, the CityLife development project sold €88 M of luxury buildings in 2 h from the sales opening) and the absence of residential buildings intended for long-term renting. This problem, if not addressed, creates the risk for

Milan to remain an isolated land closed to the outside, not attracting any more young students and workers that, until now, created the added value for the city with their knowledge and entrepreneurial spirit. A possible solution resides in the progressive expansion of Milan toward the neighboring municipalities, with the integration of them inside Milan's mechanism through transportation's connections and through smart urban developments in strategic areas at the limit between those small municipalities and the inner city.

At the same time, the opportunity of developing affordable housing solutions should be addressed by real estate stakeholders, understanding there is a market, especially in specific areas of the city, for this type of buildings (see, for example, the residential developments of Bocconi University, that offer students a low-cost alternative to renting apartments in the nearby areas, including additional services such as common spaces, sports facilities and laundry areas that improve the quality of their stay with a relatively low cost for the developers).

Finally, the growth management represents the third and last challenge faced by the city. Milan and Bologna are the only district's capitals that, among the 20 present in Italy, will grow in terms of inhabitants in the next 10 years. Despite this prediction is hardly seen as a threat for the city, Milan and its stakeholders should realize that growth has to be monitored and managed in order to take advantage of the positive benefits that arise from it and avoid that those benefits are only enjoyed by a part of the citizens.

First, the problems illustrated above, together with the ones that will for sure come up in the future, should not be ignored but addressed in a comprehensive way, with the political characters acting as facilitators, welcoming constructive feedback on the problems and finding ways to make the various actors collaborating effectively to solve them.

Secondly, Milan should avoid acting as a city-state and continue on the path of giving the example to the whole country of good management and good diffused services. The knowledge, the ambition, and the spirit of the city has to be diffused, through an open-innovation mindset, to the rest of the country, avoiding remaining isolated and penalized by the central government and at the same time contributing to the development of the other areas of the country.

Finally, the various actors in the urbanistic development of the city, from developers to financial institutions, from the municipality to local utility firms, have to gather together and find a way to align their interest toward the path of sustainability, understanding that sustainability is the only way, in the real estate environment, to do business going forward.

Chapter's Note

The content of this chapter has been crafted on the basis of a business breakfast held in Milan in February 2020, organized by Claudio Scardovi, sponsored by AlixPartners and conducted by Gianmario Verona, dean of Bocconi University and Pierfrancesco Maran, Milan councilor for urban policies, agriculture, and sustainability. The participants to the meeting have been the following: Davide Albertini Petroni, managing director at Risanamento Spa; Andrea Beltratti, professor at Bocconi University and president of Eurizon Capital S.G.R.; Alessia Bezzecchi, program director at SDA Bocconi; Armando Borghi, chief executive officer at CityLife; Stefano Bortolamei, head of digital business partner's area at Intesa Sanpaolo; Caterina Campagna Weiss, consultant at AlixPartnersM; Anna Carbonelli chief executive officer at Intesa Sanpaolo Casa; Sandrine Chapuis, marketing director at AlixPartners; Ugo Debernardi, chief executive officer at First Atlantic Capital; Paolo Dotta, co-founder at Altar.io; Cinzia Facchi, editor at Egea Spa; Federico Rigoni, chief revenue officer at TIM; Stefano Focaccia, managing director UTP at Prelios S.G.R., Nicola Forti, director of Bancaria Editrice at A.B.I. Associazione; Andrea Galati, head of foreign M&As at Intesa Sanpaolo; David Galli, head of banking services at Banca Popolare di Milano; Roberto Gamba, editor at Egea Spa.; Micol Gardoni, marketing and communications manager at Lendlease; Stefania Gentile, general manager at Mercury Payment Services; Biagio Giacalone, head of active credit portfolio steering at Intesa Sanpaolo; Flavio Gianetti, head of group's M&A at Intesa Sanpaolo; Gaudiana Giusti, independent director at A2A; Fabrizio Leandri, chief lending officer at Banca Monte dei Paschi di Siena; Rossella Leidi, general vice-director at UBI Banca; Gaetano Lepore, head of real estate—Italy at UBS Asset Management; Laura Maida, head of strategic initiatives at Intesa Sanpaolo; Piero Masera, managing director at AlixPartners; Mario Mazzoleni, member of the board at Abitare In; Stefano Minini, executive advisor at Lendlease; Claudia Parzani, regional managing partner—western Europe at Linklaters; Enrico Piazza, strategic initiatives team leader at Intesa Sanpaolo; Fabio Piccinini, partner at EY-Parthenon; Lorenzo Pietromarchi, managing director at AlixPartners; Massimo Racca, chief executive officer at Gruppo General Holding; Emanuele Ratti, head of financial services at Google; Paolo Rinaldini, managing director at AlixPartners; Francesco Rossitto, co-head of FIG—Italy and Greece at Mediobanca; Silvia Maria Rovere, president at Assoimmobiliare and chief executive officer at Morgan Stanley S.G.R; Andrea Ruckstuhl, EMEA director at Lendlease; Lucia Savarese, head of NPLs at Banca Monte dei Paschi di Siena; Riccardo Serrini, chief executive officer at Prelios Credit Services; Gina Sorce, director at ROCK Communications; Luca Tedesi, head of real estate department at Intesa Sanpaolo.

Appendix: Behind the High-Tech Curtains of a Sustainable City

Paolo Dotta and Claudio Teixeira

The city of the future is a city that revolutionizes the way its residents engage not only with the city per se (infrastructures, modern housing, transportation) but, above all, with all the activities that—day by day—affect residents' wellbeing (sixteenth-century calque of the Italian concept *benessere*) across the Maslow's pyramid: basic needs, psychological needs, self-esteem needs. In the following pages we will go through some technology advancements that can revolutionize some of the activities that, as human beings, we carry out every day in an urban environment.

SCaaS: Sustainable City as a Service and Open Data

This first use case is actually the parent of the following ones. This one makes sure that cities of the future are actually designed and managed the correct way. On top, this is responsible for many other advancements that will come in the future, since it allows the spread of knowledge and free contribution by the tech community. As much as our life lives around open source contributions in the application world, the same is happening around solutions that enhance our relationship with the urban context we are living in. New cities or city expansions are nowadays integrating a Smart City Operating System on the design phase. In the market, there are some commercial solutions for operating a sustainable city, and the motivations are to enable and enhance the city's intelligence—that can then be translated into resilience, safety, and sustainability. From urban planning, incident reporting, community contribution to real-time traffic signals auto adjustment, sustainable cities are setting a new standard of urban living.

Solutions like CitySynergy by Deloitte are usually offered as a SaaS which supports most of the city operations management and real-time telemetry visualization. City councils have now at their disposal real-time measures, correlations,

© The Author(s), under exclusive license to Springer Nature Switzerland AG 2021

C. Scardovi, *Sustainable Cities*, https://doi.org/10.1007/978-3-030-68438-9

predictions, and management suggestions of events end-to-end across the city. This information also allows mayors and their teams to make long-term decisions based on realistic and solid foundations. These commercial solutions have modules that are delivered to different audiences: Operator User Interfaces, Command Center Dashboards, Executive Dashboards and Citizen Tools (to report incidents and others). Some of the metrics collected are places, occupancy levels across indoor and outdoor public places, integrated mobility (scooters, bicycles) usage, electric chargers network usage, disaster monitoring and management (e.g., COVID War Rooms), smart meters metrics, air quality, tax payments metrics, within others.

As per the software industry, for every framework released by a tech giant, there is another framework that is created by the open-source community, namely CityOS. OS solutions might find early adopters among lower per capita cities or those that are ruled by tech-savvy councils. This enables any city in the world to begin its digital transformation toward more sustainable terms and at the same time to increase the quality of life of the residents. Solutions like CityOS follow the "open" paradigm of the software industry and allow by default residents to access data collected and contribute to their development. Software solutions like CityOS usually comply with standards, allowing sustainable cities to easily interchange data and become connected. Analogous to the commercial solutions CityOS platform brings dashboards with real-time data visualization and co-relations to be presented on multiple screens and usually deployed on the city council. Open Data initiatives have been created to allow a unified way for governments to share some data openly with the community. Data provided by governments and cities are usually licensed under CC BY 4.0 and CC0 1.0. Access to data is usually dispensed via RESTFul APIs and an API key is required in order for citizens or businesses to consume public open data. As aforementioned, access to this data is important as it allows both businesses and citizens to improve their processes and lives.

In Europe, the official regulator is the Publications Office of the European Union under the European Commission. We have been noticing more cities adopting open data strategies and stepping forward to becoming sustainable cities. Important to note that exposing some of the city information as open data is usually not a standalone initiative and belongs to a bigger plan of the digitalization of the city which is usually referred to as a sustainable city. In Europe, we can find cities offering from dozens to thousands of datasets (e.g., Lisbon, Dublin, Florence).

Edge Computing and Augmented Reality as Personal Shoppers

The Retail Industry is rapidly evolving mostly based on the successful experiments of major online retailers. Smaller and legacy large retailers are actually following the leaders by moving to and focusing on the innovation of their businesses on the communication of their offline and online channels. In a digital environment, there is one variable that makes or breaks businesses in no time: user experience. Customers are more and more demanding in terms of services that enhance their purchasing experience, such as speed and precision in the delivery, full transparency in the purchase transaction, zero waiting time: customers want it all and immediately. Retailers are hence adapting their offline and virtual points of sale in order to be able to be at the forefront of the customer experience curve.

Retail companies like IKEA have made use of advances in VR/AR to enhance further the online customer experience. VR/AR allow customers to preview in near-reality cameras (through their phones/tablets) how furniture and other goods would look in their place. Style and size of furniture and other goods can be assessed in a matter of minutes without getting out from home, driving miles to the closest IKEA mall, parking, queuing, measuring, comparing potentially making mistakes, and doing it all from the beginning. The physical shop is now (and will be more and more) a place for the "last touch" before the purchase. On the retailer side, the aim is to obviously contribute to faster sales and inherently decrease returns processing and costs. IKEA Place is an example.

Another example is brought by Farfetch, the fashion giant founded by Portuguese tech mogul Jose Neves. Farfetch coined "Augmented Retail," underlying the interaction between the virtual and physical worlds. The concept brings the magic touch of the offline shopping activity (human interaction with sales assistants), enhances it with benefits of the online experience, and ultimately manages the interaction with the customers with massive use of data. A sort of KYC performed in real-time, through advanced techniques, to be able to offer the best experience to every customer, in a personalized way. Farfetch created this retailed operating system and now it is getting traction with the luxury brand Chanel.

AR/VR solutions are generally available in modern versions of Android and iOS, and this will be growing exponentially given the consistent customer base to continuously run tests on. Applying VR/ AR on retail is only starting to emerge and there is plenty of room for improvement and spread (especially to make the various experiments viable from an economical point of view, evolving from traditional and online shopping to a "virtual hands-on" pre-buying experience). Given the amount of data that is being generated and assessed, edge computing comes in handy. Edge computing represents a shift of paradigm in computing, processing, and analyzing of data along a network edge, closest to the point of its collection so that data becomes actionable. The massive development in the IoT pushed network bandwidth requirements to the limit, sometimes creating friction in the applications' response time. With edge computing, the computer workload is closer to the consumer, and this reduces latency, bandwidth, and overhead for the centralized data center. In this specific case study, we have seen IKEA Place being distributed (Edge computing is indeed a distributed computing paradigm) to the customer's smartphones making this an inexpensive move when compared with in-situ high computing AR/VR devices.

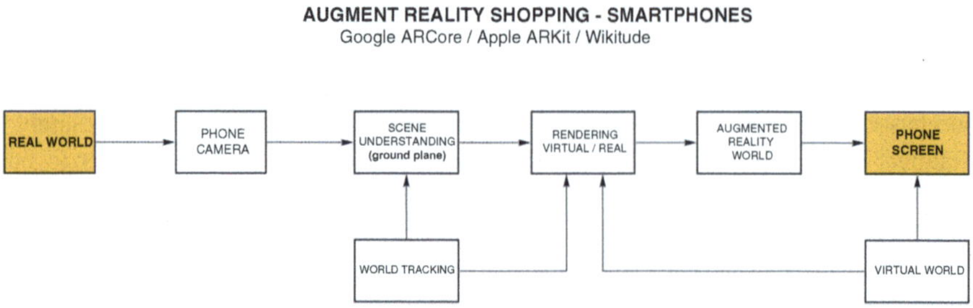

Augment reality shopping with the help of smartphones

Weather of Things to Produce Healthy Food

Weather predictions are usually based on mathematical models and are limited to the accuracy of the locations where the model runs upon. The development in IoT and edge computing enabled a superior level of weather modeling and prediction by having on-site sensors whose data is applied over the standard models to enhance and increase the accuracy of the predictions. This is indeed possible thanks to the edge computing paradigm shift. Internet of things is here defined as "weather of things," meaning a handful of devices (fixed at a specific geo point) or mobile (e.g., embedded into cars, bikes, drones) that allow to report data at a very micro level.

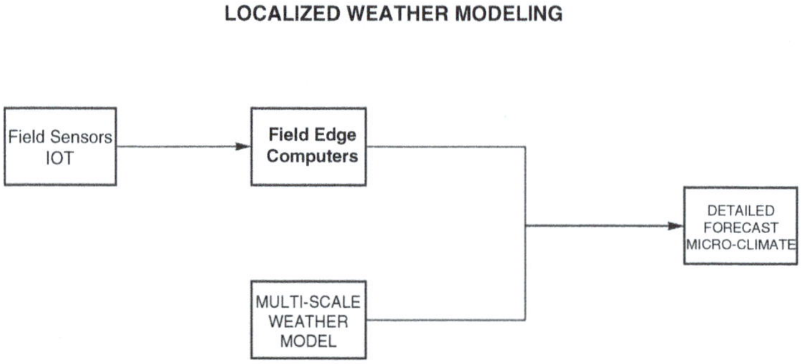

Weather of things and localized weather modeling

There is maximum granularity for each micro area: common use cases are vineyards, where microclimate models are generated, allowing better planning (pruning, harvest, other decisions to treat grapevines) and decreased business risk. These localized weather IoT devices allow also to predict with accuracy other types of weather information rather than the standard average temperature, rain, cloud coverage, pressure. Like so, values like air humidity, accurate temperature, presence of water on the ground are presented as powerful metrics for open-field entrepreneurs and workers. The aforementioned does not represent a new area, irrigation systems and mobile weather stations have been used for a while. As per the augmented retail, the innovation here is the possibility of running viable edge computing devices on a field that process and update the weather prediction models in real time. The mobile edge computing local calculation without the burden of procuring and operating a server, which could be quite challenging for rural agricultural applications or intensive vertical farms, where the management of the crops is entirely automated and delegated to sensors. Weather of things (and combination with AI) is also at the base of modern vertical farming. Given the increasing worldwide population, the massive migration toward urban centers (megalopolis), the decreasing water availability, increasing climate change issues—greatly due to farming itself, directly and indirectly—vertical farms can play a leading role in the transition toward a full, sustainable city.

Vertical farming is based on the concept that crops can be grown in a soilless environment, hence in vertical—rather than on a large, plain, field. Crops can have their roots submerged in water (hydroponic farming) or hanging in a sort of a "nutrient fog" without an aggregate medium (aeroponic farming). Aeroponic farming is being currently explored by agri-tech companies given the important role that can be played toward a sustainable environment:

- A great amount of plant growth: the nutrients reach the root with no interference from the growing medium
- No pesticides and reduce water waste: the absence of soil drastically decreases the risk of diseased reaching the plant, and the fact that there is not a liquid nutrient solution means aeroponic farming uses only 5% of the water required in traditional farming.
- Farming opportunities wherever, whenever: vertical farms are indoor; therefore, any product can be grown anywhere in the world in any season, contributing massively to the reduction of pollution made by the transportation of crops all around the world.
- Vertical space instead of horizontal space, leaving space to grow cities and facilities to accommodate the 10 billion people by 2050. Vertical farming uses 1% of the land that is required by traditional farming.

Vertical farming, though, presents one constraint to be effective and useful at an industrial level: the constant monitoring (and automated action) at a crop level is essential to understand the best configuration of the environment and nutrient variables to boost productivity at an economically viable level. And this is where the weather of things comes in help. Obviously, this "weather of things" (WoT) is not solely beneficial to the farming industry, but rather to a variegated basket of industries, such as insurance, construction, logistic, and including the defense that is responsible for citizens' safety and disaster prevention.

Being at Madison Square Garden While Being at Home

Gaming and Entertainment are relevant because, as per the aforementioned retail experience, it is another confirmation that there is a blurred line between the offline and the online worlds. We are now watching a soccer match either at San Siro or through our 4k screen in our living room, but are we sure that this is not going to change in the future? Nowadays when buying an HD television, most of us go for 4k resolution models, and 4k is indeed a pretty decent resolution. Now imagine a stage filled with a myriad of screens that result in a 32k experience (60 fps) and where all primary cameras are tracked so that data is mapped to the LED screens on stage. If we are adding Augmented Reality as well, now we can picture what has been done in Shanghai by Riot Games (well-known game developer), Possible (XR producer for Live Events, that worked with artists like Paul McCartney, Lady Gaga and Ariana Grande among others and in events like Coachella and SuperBowl), and Lux Machina (technical video solution developer): "the most elaborate mixed-reality stage in the world."

They reproduced a real environment, on-screen, 360 degrees around the stage. Michael Figge, co-founder/creative director of POSSIBLE said "So sometimes you'll see a camera person aiming their camera into the ground or aiming it back towards where our technical front of house is. And that's because there's actually a 360 world that's built out around the stage where, in any direction you point your camera, there's something to be shot." The result for people at home watching the show—through VR goggles—will exactly be the same as being at a physical location, with the same colors (almost), real supporters nearby, and along with the same vibe.

To leverage this realist to home, companies like Oculus have been working on improving their VR headsets, making them generally available and affordable with sufficient computing power. For instance, a wireless headset model is basically a high-end smartphone loaded on a case with a special screen and using a mobile operating system (Android or iOS). The magic comes from the hardware sensors (LIDAr and others) which are being processed at the edge (in the frontend!) on the smartphone and allows 3D special placing of the user with a close to reality precision. Despite these great advances, these solutions are not still commercially viable. We believe that advances in screens and the computing power of smartphones will improve these wireless and portable VR headsets.

To be so, the primary objective of a sustainable city—and its cyber real estate—is to offer all its residents and tourists a series of services to drastically increase their well- and wealth being, no matter the social status and no matter adverse situations like crisis and pandemics. We brought this use case for two reasons: the first because this tech advance can play a leading role in residents' happiness, given the exciting way they will consume content. Second, because it is a tech advancement that promotes inclusion: we are experiencing lockdown? No problem, we can still be at the arena supporting our beloved team. Cannot we afford to fly 10 h plus the entrance ticket plus the hotel? No problem again, at a fraction of the price we can still entertain ourselves on our passion through a very immersive experience (almost being in the physical place for real!).

Air Delivery and Inspection

The combination of flying drones, accurate geolocation, and edge computing makes this case one of the most important revolutions at a sustainable city level. It will have a positive impact on logistics and transportation (costs are cut up to 80% and delivery times reduced drastically), will contribute to defeating traffic jams in urban areas, to diminish accidents to goods and drivers, it will contribute to customers' happiness due to greater user experience (one of the key trait of sustainable cities) and most importantly will play a key battle in fighting climate change, both for the near-zero carbon footprint when flying around and also for acknowledging much faster adverse situations due to adverse climatic conditions. Drones have been used in commercial applications for the last 2 years with examples of TMT and logistics companies like DHL and AT&T. Drones have been used for primarily two reasons: for technical inspection and to move packages around.

The evolution of the drone technology has made it possible to have inexpensive drones with embedded edge computing devices, with computing power close to what we see in the commercial market: the improved battery packs and low weight of the drones allow faster and more accurate technical inspections on large areas. This type of use case is immediate and does not require significant upfront investments, since a commercial drone with an IoT device attached (with sensors) for taking measurements does not present any technical obstacle. The second major use case is the logistics one, and DHL is one of the pioneers in the area. A year ago, they launched a fully automated and intelligent urban drone delivery service. The company has made use of commercial drones (for example, mdMapper 1000DG from microdrones and Falcon from Ehang) which are equipped with high-resolution cameras, Georeferencing processors and remote flight execution. Most of the work done by DHL is on creating container decks for take-off and landing, attaching IoT devices for loading and off-loading of cargo boxes and software for collecting telemetry and flight planning (analogous to a flight control center of an airport). The limitation is that for now, this is only possible in urban areas due to short flight autonomy and provisioned off-loading positions. Drones are only part of the equation, as the container decks are equipped with IoT devices and edge computing to perform triage, identification, and intelligent dispatching of packets.

The manufacturer's authorised representative in the EU is Springer Nature Customer Service Centre GmbH, Europaplatz 3, 69115 Heidelberg, Germany. If you have any concerns regarding our products, please contact ProductSafety@springernature.com

Printed and bound by CPI Group (UK) Ltd, Croydon, CR0 4YY
30/09/2026
02225692-0001